Managing Virtual Teams:

Practical Techniques for High-Technology Project Managers

For a complete listing of the *Artech House Professional Development Library*, turn to the back of this book.

Managing Virtual Teams:

Practical Techniques for High-Technology Project Managers

Martha Haywood

Artech House
Boston • London

Library of Congress Cataloging-in-Publication Data

Haywood, Martha.

Managing virtual teams : practical techniques for high-technology project managers / Martha Haywood.

p. cm. — (Artech House professional development library)

Includes bibliographical references and index.

ISBN 0–89006–913-1 (alk. paper)

1. Industrial project management. 2. High technology industries. 3. Industrial project management. I. Series.

HD66.H39 1998

658.4'02—dc21 98-33847

CIP

British Library Cataloguing in Publication Data

Haywood, Martha

Managing virtual teams : practical techniques for high-technology project managers. – (Artech House professional development library)

1. Teams in the workplace 2. Industrial management

I. Title

658.4'02

ISBN 0–89006–913-1

Cover design by Lynda Fishbourne

International Standard Book Number: 0–89006–913-1

Library of Congress Catalog Card Number: 98-33847

10 9 8 7 6 5 4 3 2 1

Contents

Preface

This book provides practical tools and techniques for managers faced with the challenges of managing and leading geographically distributed teams. What do you do on the day after the merger when you wake up and find out half your team is on the other side of the country (or worse yet, on the other side of the world)? It happened to me and if it hasn't already happened to you, it is probably only a matter of time. If you are a person who is interested in theoretical discussions of the sociological and global economic implications of the virtual office, you should put this book down now. If you are one of those people who has to stand up and explain to your boss or shareholders why the project did or did not happen, this book is for you. My partners and I are practicing high-technology project managers. We have devoted the last several years to studying the best practices for managing virtual teams and applying them to the projects that we manage everyday.

In the following chapters we will address the problems most commonly identified by managers working with distributed teams. After interviewing several hundred managers, we found that most were concerned about communication, control, monitoring, and team building. Managers also expressed reservations about cultural, technological, legal, and process issues. We heard plenty of war stories and spoke to several managers who doubted that distributed teams could ever be successful. We knew from our own experience that it was possible to succeed and we found other successful managers who helped us build on that knowledge. After all was said and done, we found that none of the issues associated with distributed teams were insoluble or intolerable. In fact, managers were able to identify significant advantages. An increasing number of managers had made the transition to the new work style and were becoming evangelists. We will present case studies illustrating how successful managers created teams that were more productive, more effective, and more flexible than colocated teams.

What made it initially difficult for managers was the lack of a systematic approach for transitioning teams from their old "colocated" work style to the new "distributed" work style. Because the changes are so numerous and so pervasive, managers are confused about which changes to make, what order to make them in, and how fast to make them. Like most other types of organizational change, building effective distributed teams doesn't happen overnight. Managers and team members need to be allowed to walk before they are asked to run. At MSI we have developed a framework of principles and best practices based on the experiences of practicing managers who have both succeeded and failed in managing distributed teams. This framework, which we call the "Maturity Model for Distributed Teams," allows managers to assess the current state of their organization and put together a plan for moving their team to the next level. A good transition plan moves people, processes, and technology forward together.

The book discusses how to effectively use new technology, but the emphasis is on a set of communication principles and management practices that won't change with the next release of software. Companies need to realize that high-speed networks and groupware are tools that managers can use to solve problems. They are not solutions to problems in and of themselves. Companies often make the mistake of investing heavily in technology without making a corresponding investment in people, planning, processes, and training. If you put a spaceship in someone's driveway, you shouldn't be surprised when they don't jump in and drive it to work. As a matter of fact, they are more likely to run to their therapist to discuss how it's not really there, or call the police to make it go away. Successful managers of distributed teams set up an infrastructure of policies, processes, procedures, and training in addition to tools and technology. The technology discussions in this book are aimed at making managers informed consumers of technology, not designers.

It is not an easy job to be a manager in today's flattened, downsized world. If you've survived this long, you're a member of a tough crowd. Glossy magazine covers about "Virtual Leadership" and "Global Teams" take on a different light when you are one of the people actually responsible for being on time and on budget. This book shares the insights, research and experiences of my personal heroes, the managers who decided to swim instead of sink when all the rules changed.

1

Introduction

If you're like me, you didn't go looking for a virtual team to manage, it probably came to you. In the early 90s, I was a director of engineering at Telebit Corporation, a California-based company which made data communications equipment. I was pretty happy minding my own business, managing hardware and software developers when (without asking my permission) Telebit decided to merge with another data communications company on the east coast. Suddenly, things changed. Half the team was on the other side of the country and the problems began. I started looking for reference material on the subject of geographically distributed teams and wasn't able to find anything helpful. At the time, I was on the board of directors of the Santa Clara Valley Chapter of IEEE Engineering Management Society. I thought it would be a great idea to hold a panel discussion. I was certain that if I got together a few seasoned managers who could share their experiences, it would become clear to me how I should address the issues I was facing. I called 46 managers in Silicon Valley and had a terrible time getting anyone to agree to talk about their experiences publicly. That's when I knew this was an area with a lot of opportunity for research and improvement. Shortly thereafter, I joined several other managers to form a company devoted to researching and best practices for managing distributed teams. This book will share with you the benefit of the last three years of our research and experiences.

1.1 What is a virtual organization or distributed team?

There are almost as many answers to this question as there are people to ask. The key to answering it correctly may have to do with first understanding who is on a team. Are the people on your team only the people who report to you? Are they only the employees of the company you work for? Are they only the people in your building? Are they only the people who like and agree with you? If you work with consultants, contractors, vendors, or joint venture partners, are they members of your team? The majority of the project managers I deal with will tell you that if someone is supposed to be contributing to the outcome of the project, they're on the team. Some types of distributed team members that our clients work with include:

- Individuals located at other corporate sites
- Joint venture partners
- Telecommuters
- Consultants
- Third-party developers
- Vendors
- Suppliers
- Offshore development and manufacturing groups
- Satellite work groups
- Customers
- Clients

Some of the factors that our client's believe make their teams virtual or distributed include:

- Geographical separation of team members
- Skewed working hours
- Temporary or matrix reporting structures
- Multi-corporation or multi-organizational teams

For the purposes of this book, the primary factor that distinguishes a distributed team from other types of teams is that one or more of the team members is geographically separated from the other members.

1.2 Why are distributed teams becoming more prevalent and how rapidly is the trend growing?

I often talk to managers who think they will be able to change jobs and get away from the issue of managing remote workers and remote work groups. The reality is that you can run, but you can't hide. The implementation of distributed teams is a very rapidly growing organizational trend. Some factors driving the prevalence of distributed teams include:

- Mergers
- Acquisitions
- Downsizing
- Outsourcing
- Technology
- Clean air laws
- Offshore development and manufacturing
- Technical specialization

If you include vendors and customers as a part of your project team, almost 100% of project teams are distributed today. However, since vendors and customers have historically been off site, it's probably more interesting to look at growth in other types of distributed team members.

In the middle 90s, approximately seven million people telecommuted annually in the United States. The Gartner Group reported that by 1999, 80% of companies will have implemented some form of telecommuting for about half of their work force [1]. This doesn't mean 40% of the work force will be working from home full-time. It does mean that by the turn of the century about half of the people who are employees of your organization will be working at home part-time or participating in some form of alternative office arrangement.

Outsourcing and the use of third-party developers has moved well beyond the early adopter stage. The value of large contracts in the services and software market grew at an annual rate of 39% [2]. Forrester Research estimates that the outsourced technology market will grow 144% between 1996 and 2002 [3].

In the mid 90s the value of mergers increased at a 36% annual rate [4]. Mergers and acquisitions have become so commonplace that even if you've managed to stay in the same job for a number of years, the name and structure of the organization you work for has probably changed. One of my clients has been sitting in the same office for four years doing the same job. During that time his business card has had three different company names on it.

Corporate downsizing has caused many managers to look at the use of contractors, consultants, and temporary personnel as a means of dealing with project staffing requirements. ATT has increased its use of consultants at a rate of 100% per year [5]. Gartner Group predicts that by the year 2000, more than 80% of large enterprises will *routinely* use consultants [6].

New communications technology is one of the primary enabling factors for distributed teams. As the cost of bandwidth decreases, the return on investment for distributed team members skyrockets. In 1995, the data rate for standard phone lines was 14.4 kbps. In most areas of the United States the cost for this type of standard residential phone service was about $12 per month. Today, almost every regional Bell Operating Company has announced ISDN tariffs in the range of

$24-$25/month. ISDN service provides a low-cost, high-bandwidth phone line (128 kbps) for both residences and businesses. For only twice the cost, a TENFOLD increase in performance is available. Cable modems and XDSL technology promise to continue to push this price/performance ratio. By the turn of the century we should see very low-cost multi-megabit rates to the home and office. Consumer demand for bandwidth seems almost insatiable. I suppose you can never be too rich, too thin, or have too much bandwidth.

To return to the question posed at the beginning of this section, "How rapidly is this trend growing?" Attaching a single number to the trend is difficult since each of the factors driving the prevalence of distributed teams is growing at double or triple-digit rates. If you are a high-technology project manager, managing distributed teams is already a required job skill.

1.3 What are some of the advantages of a distributed team?

One of the exercises we do in our seminars is to ask the attendees to create two prioritized lists. One list contains their perception of the advantages of a distributed team. The second list contains their perception of the challenges of a distributed team. We collected results of these exercises and found that they were very different depending on the composition of the audience. Executives and managers tended to have a completely different perspective from team members. As all project managers know, when the managers and the team members have different objectives something needs to change. The next few sections should give you an idea of what managers expect, what team members expect, and what research says actually happens.

1.3.1 The managers' perspective of the advantages of a distributed team

Looking at the lists that managers generated, 98% of the time the item at the top of list had something to do with cost. Eighty percent of the time *everything* on the list had to do with cost. A typical list might look like this:

- Access to a less expensive labor pool
- Reduced office space
- Greater utilization of employees
- Round-the-clock work force
- Greater access to technical experts
- Larger pool of possible job candidates

1.3.2 The team members' perspective of the advantages of a distributed team

It may or may not surprise you to find out that the team members had a completely different perspective. Ninety percent of the time the item at the top of their list had nothing to do with cost. As a matter fact about 65% of the time nothing on their list has anything to do with cost. Team members are more excited about increased independence and greater job opportunity. A typical list might look like this:

- Increased independence, less micro management
- Larger pool of jobs to choose from
- Greater flexibility
- Opportunity for travel

1.3.3 Matching expectations to research

In order for your virtual organization to be successful, both the team members and the managers need to understand how the organizational change is benefiting *all the participants.* Although most executives and managers are motivated by cost-oriented objectives, research done by Becker and Steele at the International Workplace Studies Program at Cornell indicates that alternative office arrangements that are strictly cost driven *almost always fail* [7]. Alternative or virtual office arrangements are successful when team members understand how the changes improve overall business process. Successful managers of virtual teams understand the team members' objectives and expectations and incorporate them into the transition plan.

Case studies of distributed teams demonstrate an impressive list of advantages, some of which are not immediately obvious. Our research found documentation supporting the following:

Increased productivity

Numerous studies confirm that telecommuters show productivity increases between 15% and 80% with an average increase of 25% to 35% [8]. Our studies indicate that productivity increases are not limited to work-at-home situations. Individuals working on projects headquartered at geographically remote corporate sites also reported productivity increases due to more structured, less interrupt-driven communication styles.

Improved disaster recovery capabilities

Oracle Corporation's customer support division recently implemented a plan using virtual teams to reduce susceptibility to environmental disaster. By geographically distributing their support analysts, the company will be able to continue to provide round-the-clock support if there is a earthquake in the Bay Area or a blizzard in Colorado.

Increased employee satisfaction and retention

The results of the Smart Valley Project (a study on telecommuting in Silicon Valley) showed a 24% in employee satisfaction [9].

Reduced office space requirements

A recent study of major American corporations conducted by Cornell University found that on an average day, 40% of desks are unoccupied [10]. IBM successfully implemented a hoteling program for its marketing and services personnel which resulted in a 60% decline in real estate costs per location [11].

Environmental benefits

The 120 million US commuters produce 38.5 million gallons of exhaust fumes each year. Studies in the United Kingdom showed that taking one commuter off the road eliminated 1.2 tons of carbon dioxide [12].

Closer proximity to customers

In 1995, Xerox closed its Waltham, Massachusetts office and sent 5,550 sales and support personnel to work at customer sites [13]. Xerox reported that increasing product complexity required employees to spend more time working at client sites.

Increased flexibility

Nagel and Goldman's book, *Agile Competitors and Virtual Organization* cites numerous examples of how groups of small companies form "networked organizations" that outmaneuver and outperform the Fortune 500.

Greater access to technical experts and a larger pool of potential job candidates

In our consulting practice we see clients increasingly recruiting team members from a global market. The offshore software development industry is expanding at an astounding pace.

1.4 What are some of the challenges of a distributed team?

Managers and team members also have different views regarding the challenges of distributed teams. One of Steven Covey's "Seven Habits of Highly Effective People," is to "Seek first to understand, then to be understood." The point of comparing managers' concerns to team members' concerns is to help prepare managers to lead organizational change. If managers do not spend time up front addressing the fears and challenges of team members it is impossible for them to get the cooperation necessary to institute the reporting and control mechanisms managers want and need. Team members can be paralyzed by being put into a new situation without the necessary training and support structures.

1.4.1 The managers' perspective of the challenges of distributed teams

When managers created prioritized lists of what they perceived as the challenges of working with a distributed team, about 70% of the time the number one item had to do with control. "How do I monitor performance?" "How do I know people are working on the right thing?" "How do I train and mentor new employees?" Communication was a close second. Many times the entire list consists of variations on these two themes. (Chapters 2 and 4 address the issues of communication and control) Managers are also concerned about:

- Team building
- Cultural issues
- Cost and complexity of technology
- Process and workflow

1.4.2 The team members' perspective of the challenges of a distributed team

A typical list from team members looks like this:

- Communication
- Technical support
- Recognition
- Inclusion vs. isolation
- Management resistance

Team members are most concerned about communication and support. For non-technical team members, the issue of being isolated from technical support is overwhelming. This is not limited to satellite office or work-at-home situations. Often an individual working at a remote corporate site has no options for certain types of local technical support. In our consulting practice we have often seen situations where the local system administrator makes sure that the local desktop hardware stays up, but he disavows all knowledge of remotely hosted applications and remote access equipment. One remote team member working on a Internet-enabled application compared her job experience to going to the movies and being forced to watch the projector. Although she felt working in a distributed team had great potential, she saw all of her productivity gains eaten up by tools problems. Successful distributed teams have a plan for providing the same level of technical support to remote team members that they provide to local team members.

Team members are also concerned about being cut off from informal team communication. Many team members cite the "out of sight, out of mind" phenomena. They are afraid that team members at the central site will find it easy to exclude them from key meetings and decisions. There are frequent discussions of how hard it is to get remote team members to return a phone call or an e-mail. (Chapter 2 discusses how managers can create a communications infrastructure which addresses these issues.)

Team members are worried that managers won't realize how hard they are working. They are concerned that all of the promotions and good job assignments will go to the team members who work at the same site as the manager. Managers need to assure remote team members that they will be evaluated based on clearly understood performance objectives. (Chapter 4 discusses techniques for increasing visibility and developing performance objectives)

It is only natural for managers to want to focus on status reporting and control mechanisms because these are the issues that strike fear in their hearts. In our studies we have found managers will not be successful in addressing their control issues until they address the team members' support and communication issues.

1.5 How do managers really feel about this?

Between July 1995 and January 1997 we surveyed 514 managers at high-technology companies. We conducted the survey to help us understand managers' views and attitudes about distributed teams. The actual survey questions and responses are included in Appendix E. In reviewing the data, it is important to realize that the data was collected from managers who were interested enough to attend a course, trade show, or lecture on distributed teams.

What the survey told us was that managers had a lot of reservations about virtual teams. The overwhelming majority of managers found geographically distributed teams more difficult to manage than colocated teams. They felt that projects using geographically distributed teams take longer to complete and that separating team members results in serious communication problems. When given a choice between a remote worker and colocated worker most managers were inclined to hire the colocated worker.

For the majority of managers, working with virtual teams is a new experience. The management skills they've developed don't translate well to the virtual world. Most companies have not done an adequate job preparing managers for the shift in work style. Early adopters often feel the sting of entering the virtual workplace unprepared. Our objective in this book is to provide managers with the tools and skills they need to do it right the first time.

References

[1] Masud, Sam, "Telecommuting: Is it for Everyone? It Might Be in 1999," *Government Computer News*, Vol. 14, 3 July 1995, p. 41.

[2] *Datamation* , Vol. 41, no. 5, 15 March 1995, p. 32.

[3] Meringer, Julie,"Sizing Technology Services," *Computing Strategies*, Vol. 14, no 8, June 1997.

[4] *BusinessWeek*, 1 May 1995.

[5] *BusinessWeek* , 25 July 1994.

[6] http: / / www.gartner.com / hotc / esp0497.html

[7] Becker, Franklin, Steele, and Fritz, *Workplace by Design*, San Francisco, CA, Jossey-Bass, 1995.

[8] Katz, "Management Control and Evaluation of a Telecommuting Project: A Case Study", *Information and Management*, Vol. 13, pp. 179-190.

[9] Smart Valley Project Final Pilot Results, 25 October 1995, pp. 11,19.

[10] Gray, Hodson, and Gordon, *Teleworking Explained*, John Wiley and Sons, 1993, pp. 275.

[11] Durkin, Tom, "Overcoming Management Resistance: The Problem is the Solution at IBM," *Telecommuting Review,* Vol. 12, no. 4, April 1995.

[12] Gray, Hodson, and Gordon, *Teleworking Explained*, John Wiley and Sons, 1993, pp. 22.

[13] Nadeau, Michael, "Not Lost in Space," *Byte*, Vol. 20, no. 6, p. 50, June 1995.

2

Solving the Communications Problem

A lot of what we learn about communicating, we learn before we are two years old. Your verbal communication style is one of the sets of behaviors you've been practicing the longest. For this reason, many people never question the techniques they use for communicating and have a great deal of difficulty changing. Their communication practices are so ingrained and instinctive that they can't conceive of different methods. However, waiting for someone to look you in the eye and nod their head is not the only way to determine that you are understood. It's what my mother taught me in 1958 but, let's face it, times have changed.

Communication styles and norms are constantly evolving. I can remember dialing a friend's telephone in 1977 and getting an answering machine for the first time. I was frustrated and angry. As I stammered into the telephone, I couldn't think of what to say. I was annoyed that they were not home to take my call. Twenty years later, if I call someone who doesn't have voice mail or an answering machine, I'm annoyed. I think it is extremely discourteous of them to force me to dial their number more than once. Half the time, I'm disappointed if they actually pick up the phone, because I'm in a hurry and I just want to leave a message. Looking back, I can't identify the point in time where I became so comfortable and dependent on voice mail. As we all become more practiced at communicating at a distance, different norms are developing. These norms help us become more effective.

One source of friction among distributed team members has to do with varying levels of familiarity with electronic communication etiquette. At a recent seminar I conducted for the Project Management Institute, one of the attendees complained bitterly that his European team members were much less effective at using voice mail than his American team members. He had concluded that the British were just poor communicators because they often left messages simply asking him to call them back, instead of leaving a detailed description of why they were calling. When I inquired whether his European team members had regularly used voice mail before, or whether they had been given any training, his answer to both questions was no. Voice mail is much less prevalent outside the United States and Canada. It doesn't matter what country a person is from, if you ask someone to do something they haven't done before and don't give them any training, they're probably not going to do it the way you want it. It is so easy to take the skills you've already developed for granted. Don't assume all your team members are on the same level.

In the late 70's, Thomas Allen, of MIT's media lab, published the results of a study he had conducted to determine the relationship between the frequency of communication between coworkers and their separation distance (Figure 2.1). He found that once people got more than ten meters apart, the likelihood of their communicating at least once a week dropped to below 5%. The "old style" of communicating is greatly effected by proximity.

Our survey showed that, in the late 90's, high technology project managers were significantly changing this standard. More than eighty percent of these managers reported that they communicate with

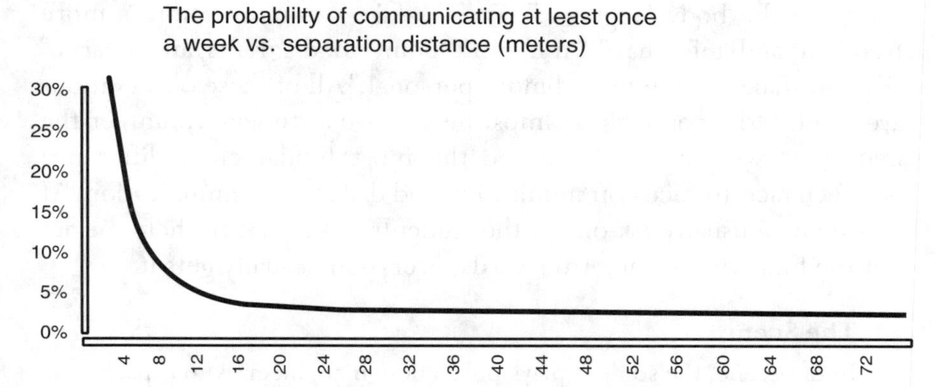

Figure 2.1 From Allen, *Managing the Flow of Technology,* MIT Press, 1977.

remote team members at least once a week. Unfortunately, they were quick to point out that communication patterns between their individual team members looked more like Thomas Allen's study. Project managers are the "early adopters" of the new communications styles because they are the ones saddled with the responsibility for making things work.

In this chapter, we will explore techniques which have been successfully used for improving communication among distributed team members. These techniques are based on four principles that we have determined are the keys to effectively communicating at a distance. Section 2.1 gives a brief description of each of the principles, and Sections 2.2 and 2.3 give examples of how the principles can be applied to various forms of electronic communication. It is important to distinguish between applications of the principles and the principles themselves. E-mail technology, groupware and videoconferencing are changing everyday. You will need to adapt your implementation techniques as technology changes. However, the principles will remain the same regardless of which e-mail package or software release you use. To understand the principles, it is good idea to spend a little time thinking about what makes communicating at a distance different from face-to-face communication.

When I ask students in our seminars what they think is different about communicating at a distance, I usually hear answers like: "You can't see the body language," "Face-to-face communication is more frequent and informal," "You can't point to diagrams and charts," "Face-to-face is warmer and more personal." All of these observations are valid. Oddly enough, I almost never have a student volunteer the issue that we have found to be the most fundamental difference between face-to-face communication and distance communication. At that point, I usually ask one of the students in the class to help me act out the following scene. Afterwards, everybody usually gets it.

The Scene

In this scene, the student plays Bob, a design engineer, who is packing his briefcase getting ready to go to the airport. He's going to deliver a paper at a technology conference. I play the customer support engineer.

Martha: Listen Bob, I need to know what error code AE277 is for the old C110 product. I have a customer whose system is down and I need to get back to them. Just take a look at this printout.
(I shove the printout under his nose)

Bob: Martha, I'm getting ready to leave. Why don't you do talk to the sustaining engineering department.
(Bob closes his briefcase and starts walking towards the door. I follow Bob closely, continuing to wave the printout in front of his face and point to the error code)

Martha: I've already been there. They don't have any documentation, and they said you were the last one to work on this.
(Bob continues to move towards the door)

Bob: Look, Martha, send me an e-mail.
(I throw myself in front of door spreading my arms to block Bob's exit.)

Martha: You never return my e-mails! I'm not letting you out of this room until you spend 2 minutes looking at this printout.

Bob: Alright, 2 minutes.

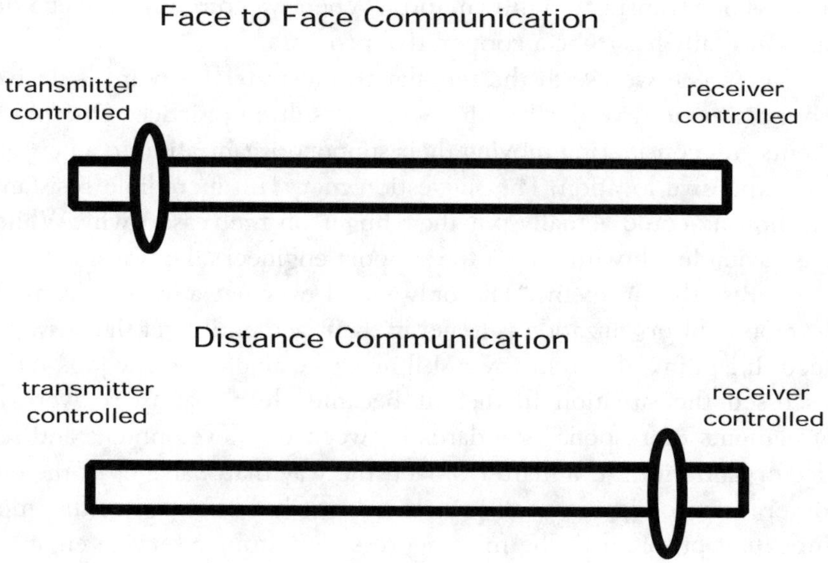

Face to Face Communication

transmitter controlled

receiver controlled

Distance Communication

transmitter controlled

receiver controlled

Figure 2.2 Comparison of control levels.

The scene illustrates one of the most fundamental differences about communicating at a distance. The issue is about who has control of the communication (Figure 2.2). Face-to-face communication is almost entirely transmitter controlled. On the other hand, distance communication is almost entirely receiver controlled. If I had called Bob on the phone, he might not have answered or he could have easily hung up.

In our culture, it is interesting how many people think that communication is something that you do *to someone* as opposed to *with someone*. I call it *assault communication. The New Century Dictionary* defines communication as "transmission…the imparting or interchange of thought, opinions, or information by speech, writing, or signs." I think it says something about the way we think about communication when even the dictionary uses the word "transmission" as a synonym for communication. In reality, our communication can't be successful unless the receiver acknowledges, understands and acts on the information. When we are in the same room with someone, we can do a lot to force an unwilling or uncooperative receiver to process our transmitted information. When we communicate at a distance it really has to be a cooperative process.

The scene we use in the seminar to illustrate this point is derived from a situation we dealt with in our consulting practice. One of our clients was considering moving their support organization to a new and less expensive location. The suggestion met with incredible resistance and no one could actually put their finger on the reason why. While I was having lunch with one of the support engineers, he was finally able to express the problem. "The only way I ever get any help from the development organization is to get in their face until I get the answers I need. If I'm in a different town, I'll never be able to do my job." As we discussed the situation further, it became clear that there were no agreements or response standards between the development and service organization. As a matter of fact, the way that manager of the customer service department prioritized work was to give the most important problem to the most aggressive customer service engineer. Distributed teams can't use this primitive method for prioritizing work and setting up response levels. An important part of the manager's job is to setup the appropriate procedural and organizational infrastructure for team members. Team members get desperate and resort to "assault communications" when they have no idea of when and how they will get a response to their communication. Section 2.1.1 gives some practical techniques for setting up availability and response standards.

2.1 What are the four key principles for effectively communicating at a distance?

As we have studied distributed teams over the past several years, we saw managers and team members trying all types of new communication practices. We looked for cases where information not only was transmitted across distances but also received, acknowledged, understood, and acted upon. We found that teams which were communicating successfully had developed practices which implemented the following four principles in some form.

1. Standards for availability and acknowledgement were defined and respected;

2. The team members replaced lost context in their communication;
3. The team members regularly used synchronous communication;
4. Senders took responsibility for prioritizing communication.

Each of these principles is explained in the following paragraphs (2.1.1 -2.1.4). Section 2.2 gives specific examples of how these principles can be applied to specific forms of electronic communication such as e-mail or video conferencing. It is important to understand that the principles for effective distance communication are *independent* of the communication technology. Once team members understand these principles, they will be able to easily adapt to a new communication tool or software release. Without a good grasp of these underlying principles, each change in technology will be a setback for your team members.

2.1.1 Developing standards for availability and acknowledgment

When you develop availability standards, you are creating agreements between team members about when and how they will be available for collaboration as well as how quickly they will respond to requests. If you want to do one thing that will dramatically improve your team's communication, you should develop availability standards. Each team member makes it known what their normal working hours are and how often they check their voice mail, e-mail, and interoffice mail. They also establish some standard for how quickly they will respond or acknowledge each type of communication. The type of availability standards that are appropriate for a team member will vary, based on that person's or group's job function. A customer support engineer may need to maintain a very high standard of availability. A research scientist may serve his team members and customers well if he responds to communication within a week. Availability standards should always be written into contracts when you are dealing with multiorganizational teams. One technique that we have used successfully for implementing availability standards within companies involves publishing availability standards on personal web pages. Figure 2.3 shows an example of a personal web page for a design engineer.

The system of personal web pages works best when all the individual pages are linked back to a department or team web page. The web page technique is merely one implementation of the principle of maintaining availability standards. You can implement availability standards with whatever tools or standards work for your team. If you don't have an intranet, you can put the availability standards into a pamphlet and mail them to team members. It's the mutual understanding that's important, not the technology. It's also important not to overdo availability. Think about what is really appropriate and necessary. With pagers and cell phones, you can make distributed team members more available than their colocated counterparts.

The other enormous benefit of availability standards is that they provide a foundation for establishing trust. In Chapter 3 we discuss how being able to rely on a specific standard for response eliminates distrust and the need for the kind of assault communication discussed at the introduction of this chapter.

The implementation of acknowledgement is more complex when we are communicating with someone remotely. When we are in the same room with someone, the process happens so quickly that we don't think of it as having the discrete steps shown in Figure 2.4.

When we are colocated, we use observation to ensure our messages are not only received, but also understood and acted upon. To communicate remotely, we may need to specifically request a distinct acknowledgement at each step of the communication process. Sometimes the information can be acted on so quickly that we only need a level 3 acknowledgement. But many times that is not true. If we haven't been systematic about the acknowledgement process we have no idea if and when our communication has broken down. Because different forms of electronic communication automatically provide different levels of acknowledgement, setting up the right acknowledgement is not an intuitive process. If I talk to someone on the phone, I may be able to determine that my message was both received and understood. If I send someone a fax, even if I get an acknowledgement from far end machine, it is not clear to me that it has reached the recipient's desktop. I need to include in my fax a request

Welcome to Joe Person's Page

Contact Information:

Telephone: (415) 555-1111
Best time call-after 4:00PM

Voice Mail:
Checked at a least twice a day - acknowledged in 24 hrs

Fax: (415) 555-4074
Delivered daily by interoffice mail - acknowledged in 48 hrs

Cellular phone: (415) 555-1803
Operable only when I'm in the car

Pager: (415) 555-2222 PIN # 12345
Answered within 1 hour

Home Phone
For emergency use only

Email:jperson@mgtstrat.com
Checked morning and afternoon - acknowledged in 24 hrs

Mailing Address: *2966 Diamond Street #243, San Francisco, Ca 94131*

Normal Working Days/Hours

9:00 AM - 6:00PM Pacific Standard Time
Monday through Friday

Tools/Applications

Pentium PC running Windows 95
Sun Sparc Station running Open Windows
Applications: Microsoft Word 4.0, Framemaker, Microsoft Excel 3.0

This Week's Status Report

Personal Info

Figure 2.3 An example of a personal web page.

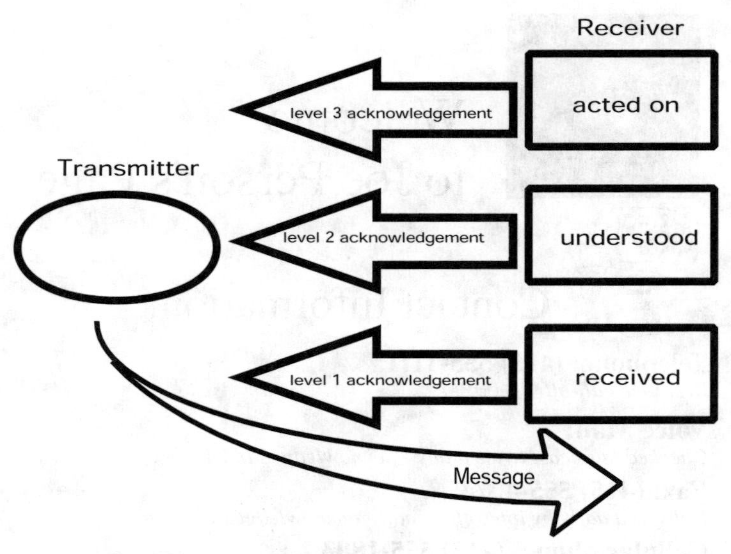

Figure 2.4 Communication model.

for acknowledgement. "Please leave me voice mail tomorrow to let me know whether you received this and whether you agree with my proposal." At MSI we call this creating "Closed Loop" communications. If you have ever studied analog electronic design, you know that "open loop" systems go out of control. If you look at an organization as system, the same theory applies. I have worked with many engineers who have very solid backgrounds in communications theory. They would never dream of designing an analog circuit without a feedback loop or writing a software communications protocol without acknowledgment and error recovery mechanisms. However, the same people will broadcast e-mails which contain critical project information but have no indication of if, when, or how the information is supposed to be acknowledged or processed. Then they wonder why team communication is so poor.

Sometimes, managers tell me "I think acknowledgement and availability standards sound like a great idea in theory, but my team members won't respect the standards." Changing group norms and communication standards usually isn't quick and easy, but that doesn't mean you shouldn't do it. You will have to do some educating and selling. There will probably be frustration and some initial resistance involved. If you have been a project manager for any length of time, resistance and frustration are probably your oldest and dearest friends. I always try to find a few evangelists in an organization. Once a few people are regularly practicing the principles and reaping the benefits, it's amazing how peer pressure becomes a force in establishing and maintaining the new norm. Think about some of the existing communications norms that are rigorously enforced by peer pressure. For example, most people will not interrupt a scheduled meeting in a closed conference room unless it is really important. Once the new norm is established, it is self-maintaining.

The case study below shows how availability standards enabled one team member to act as a mentor and train another team member from across the country. I like this study because managers often consider mentoring and training the most difficult functions to accomplish remotely. In this case, the team members established the standards themselves. By studying many of these individual instances of success we were able to abstract some general principles.

Case Study

Liz Wilcox is technical support analyst with Oracle Corporation. She works in the Colorado Springs area. Liz was assigned to mentor Rene Bastine, a new technical support analyst who was working in New Hampshire. Liz and Rene had not met before their mentoring relationship began. Both Liz and Rene had formerly worked at DEC and shared a common vocabulary in terms of both technology and work processes. Liz was not sure what subjects Rene needed help with, but she made she sure she was highly available to Rene to answer questions. Liz had two phone lines. If Liz taking a call when Rene phoned, Liz would put the current call momentarily on hold and setup a time to get back to Rene. There was no formal training plan or work process manual to help Liz assess or train Rene. Liz knew, however, that Rene had many years of experience working as a technical support analyst at DEC and quickly ascertained that Rene was very technically competent. Liz and Rene talked by phone more than once a day every day for the first 2 weeks and exchanged e-mail several times a week. Liz found that the majority of her time was spent mentoring Rene on work processes and tools. Some of the work processes that Liz walked Rene through included:

1. How to submit a bug report.
2. Where to look online for information about a problem.
3. How to order a patch.
4. How to order a software update.
5. When and how to enter notes in the notes file.
6. How to escalate a problem.

After about three weeks Rene was operating autonomously. Liz and Rene are great friends and established a very satisfying personal and professional relationship before ever meeting face-to-face.

In this case study it is interesting to see how high availability overcame several other significant negatives (such as lack of a specific mentoring plan and lack of written documentation).

2.1.2 Replacing context

In order for us to understand information that is transmitted to us, we need to have a good idea of the sender's frame of reference or context. Fussel and Benimoff have done extensive research in the area of interactive electronic communication. Their general definition of context is

> "...all speakers and hearers, attempt to construct a shared communicative context in which their messages can be produced and understood...the participants in a conversation strive for a shared understanding of the situation, of the task, and of one another's background knowledge, expectations, beliefs, attitudes..."[1]

When improperly used, many forms of electronic communication can reduce, eliminate or distort context. By consciously building context, we greatly enhance the receiver's ability to understand our message. At MSI we have found it useful to look at three different types of context:

- Physical Context
- Social Context
- Situational Context

If we share the same physical context, then I know what is within your field of view. If we have the same physical context, then I can hold up an object and use words like "this" or "that" and you will understand what I'm talking about. If we are on the phone and I am talking about a diagram in front of me that you can't see, I have to do some significant talking to verbally replace the physical context.

Social context has to do with understanding the type of person that I'm communicating with. In our society, there are sets of rules which govern how people in different social groups interact with one another. For example, if I am sitting at a stop light and a little old lady is hobbling in the middle of the street after the light turns green, social context tells me it is not okay for me to lay on the horn and yell "MOVE IT OR LOSE IT." However, if I am sitting at the same light and I see an obnoxious teenager standing in the middle of the street holding a boom box talking to one of his buddies, social context tells me that it is okay to honk and yell (that is, if I think he's not armed).

Situational context means that I have an understanding of events surrounding the person or issue. For example, if I see the boss shouting angrily at one of my coworkers in hall, I know that it is not a good time to go and ask him for a favor. I know this because we have "shared situational context."

In Section 2.2 we give suggestions of how to build context with several of the more commonly used forms of electronic communication.

2.1.3 Using synchronous communication

Face-to-face meetings, phone calls and videoconferences are examples of synchronous communication. All the parties involved are engaged in the communication simultaneously. We call this communicating in the same "time space" (Figure 2.5). When we communicate asynchronously, there is a significant delay between the time the

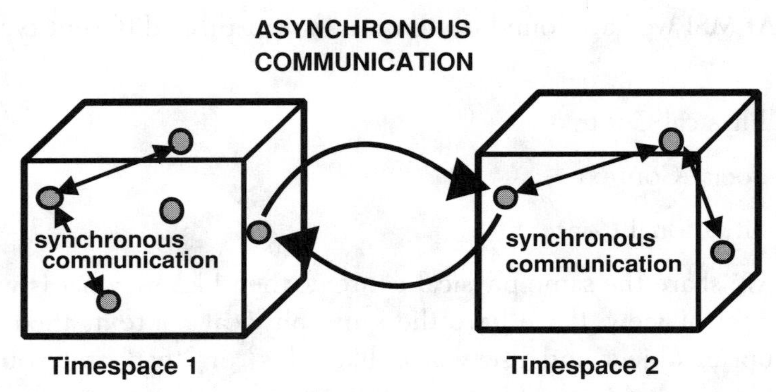

Figure 2.5 Synchronous and asynchronous communication.

sender transmits a message and the receiver receives it. Examples of asynchronous communication are e-mail, voice mail, fax, interoffice mail and web publishing.

Why is synchronous communication important? Because research supports the fact that synchronous communication tends to build relationships more quickly than asynchronous communication. In studying pairs of remote team members who established satisfying trusting relationships before meeting face-to-face, we found that the time required to establish the relationship was inversely proportional to frequency and length of synchronous communication sessions. It is certainly possible to establish relationships based strictly on asynchronous communication, it just takes longer. Dr.'s Isaacs and Tang of the Sun Center for Collaborative Computing have published several papers which suggest that the benefit of video conferencing is not in the individual meeting outcome but in the work process or social and relationship building qualities of the medium [2]. Relationships are important because we treat the people we have relationships with differently than the people we view as anonymous.

An amusing story that I think illustrates this point very well comes from an old episode of the *Twilight Zone*. In this episode called "The Button" a man gives a box with a button on the top to a woman and her husband. He explains that all they have to do to receive a million dollars is to push the button on the top of the box. As soon as they push the button, someone whom they have never met and whom they have no relationship with whatsoever, will die. The woman and her husband need the money. After a few moments of discussion, they are able to convince themselves that the person who will die is probably an old sick person and that they would be doing that person a favor. So, they push the button and the man returns and with the million dollars. The couple is overjoyed, but they are puzzled when the man asks for the box back. The wife asks the man worriedly "What are you going to do with it?" The couple recoils in terror when the man replies "I'm going to give it to someone you don't know."

If we build relationships with our team members, they are less likely to push "the button" on us. It is a lot harder to ignore someone who has personified themselves to us, somebody who has thoughts, needs, feelings and a boss that might get mad at them if they don't get their work done. I know I'm more likely to respond to an e-mail I receive from my friend John than I am to respond to an e-mail from stranger@obscure.com. Since we look at communication as being complete only when information has been transmitted, received, acknowledged, understood, and acted upon; doing everything that we can to make sure the information is actually acted upon is crucial. Having a relationship with the receiver is the best thing you can do to make sure your message is acted upon. We recommend to our clients that they establish communication patterns among their distributed team members, which allow synchronous communication at least once a week for period of two months. We sometimes give project managers a daytimer page like the one shown in Figure 2.6, to help them be more systematic in building relationships. The manager lists his or her team members in the left hand column. If the manager engages in some form of synchronous communication with that team member during the week, he checks off the box corresponding to that week. If at the end of two months he surveys the row corresponding to a specific team member and notices there are very few check marks, he knows he is not doing a very good job of building a relationship with that team member.

In my classes I'm often asked questions like: "What's the minimum amount of time I can spend in synchronous communication with each team member and still establish a strong relationship?" or "If I get all ten of my direct reports on a conference call, can I check off all ten of

Team Builder										
Team Member Name	Week									
	1	2	3	4	5	6	7	8	9	10

Figure 2.6 Team builder.

them?" There are so many variables involved and the area of research is so new that it is only possible to give general guidelines. There is also debate about what constitutes a relationship or trust. In a study that we are currently conducting, we asked a group of customer support representatives how they would know if they trusted one of their coworkers or customers. This particular group provides level 3 technical support (i.e. they handle only the most complex problems) for a company that distributes operating system software. The answer we got was "Only if I really trust someone will I allow them to dial into my machine and login with root access." When we looked at the call logs and e-mail records, there was a very high correlation between the amount of time they had spent on the phone with a customer and the likelihood that the customer would be trusted enough to be allowed dial-in root access. Customers with whom the analysts had had only e-mail contact were almost never "trusted." The amount of time required to establish trust seems to vary significantly from analyst to analyst. Although this study is still in progress, I doubt that we will be able to come up with a magic number of minutes, even with this very constrained definition of trust in this very controlled environment.

Our interviews with managers of international teams also confirmed the value of synchronous communication. Sri Chaganty is a Vice President of Engineering at Holontech. He has many years of experience using software development groups in the Philippines to work with his software development groups in California. He found a factor that differentiated successful from unsuccessful projects in his organization was presence of a liaison who spent several hours a day on the phone with one or more team members in the Philippines. His advice to other managers in the same situation was "If you don't have the budget for the liaison, you shouldn't use an overseas group." When time zones or team topology make it difficult, if not impossible, for all team members to utilize synchronous communication, designated liaisons can help teams derive many of the relationship building benefits of synchronous communication.

Case Study- IDT/Centaur Corp

IDT is semiconductor manufacturer located in California. Centaur, located in Austin, Texas, is a wholly owned subsidiary of IDT which designs Pentium class microprocessors. In order to be successful, team members at both the parent and the subsidiary company needed not only to work closely on fabrication and process issues but also to achieve technical breakthroughs in several areas. The executives at Centaur recognized that poor communication was threatening the project's success. The solution to the problem was to hire Twila Hamilton, a specialist in device physics and modeling to act as liaison to the California parent company. Twila had a very high degree of technical knowledge in both the areas of semiconductor processes and microprocessor design. Her qualifications made her credible to both local and remote team members and allowed her to make sure that information was not only exchanged but also understood. However, since Twila was operating as an influence manager in this environment, no amount of technical competency could ensure that the information was acted upon. She made extensive use of synchronous communication to establish relationships with California team members. Twila made frequent trips to California and participated in frequent phone calls and phone conferences with team members at IDT. Within three weeks she developed a particularly strong relationship with a process engineer in California. Within four months, the time required to resolve open issues between the divisions dropped by half. In this case, synchronous communication led to relationships and relationships led to action.

2.1.4 Prioritizing communication

Many of the new forms of electronic communication give senders a previously unprecedented capacity for broadcasting messages to a wide audience. One of the most common complaints about electronic communication is the resulting "information overload." The final principal for effective distance communication is that senders must take the responsibility for prioritizing communications. When team members are subjected to uncontrolled unprioritized streams of input, their effectiveness plummets. Managers need to help team members come up with a common understanding of how to prioritize the communications media that the team uses and how to prioritize messages within the media. Section 2.2 gives specific techniques for prioritizing messages within each of the different media. This section should provide some background about how to effectively prioritize the various communications media. Different team members have very different ideas about the priority of various communication media. We recently did a survey of project mangers to determine what priority they gave various communication media. They were asked to rank in order eleven different communication modes, giving a rank of 1 to the mode they gave the highest priority, and a rank of 11 to the mode they gave the lowest priority. One hundred sixty valid surveys were returned. The results are illustrated in Figure 2.7.

There is a high level of agreement about the high priority of scheduled face-to-face meetings and scheduled phone calls. Almost 59% of the respondants gave scheduled face-to-face meetings the highest priority rating. Sadly, there is also a high level of agreement about the low priority of electronically published information. Forty-eight percent of respondents gave electronically published information the lowest priority rating. With all the billions of dollars spent on intranets, a lot of executives will be disappointed to find out that without specific training and intervention their employees give this information lower priority than the interoffice junk mail. What is more interesting is the lack of agreement about the priority of e-mail, voice mail, pagers, impromptu meetings, and impromptu phone calls.

Notice how the responses are more evenly spread across the priority ratings towards the middle of the chart. The respondents to this survey were a relatively homogeneous group, project managers and project management students. Although we have not conducted the survey formally on a more diverse group, I strongly suspect that the members of a cross-functional team would exhibit an even lower level of agreement. When team members do not have a common understanding of the priority of a communications media, messages start to fall through the cracks, team communication disintegrates, and distrust sets in. Your highest priority form of communication needs to be a media to which everyone has equal and immediate access. Our "old style" of communication said the highest priority messages were conveyed face-to-face. For distributed teams, that standard must change. Because I deal with a lot of teams which have short life spans and are multi-organizational, I find using voice mail as the highest priority works best for me. Voice mail is generally available to all team members and it can be retrieved remotely. Team members must agree to check voice mail several times a day or have their voice mail connected to a

priority rating 1=highest 11=lowest	High Priority 1's	2's	3's	4's	5's	6's	7's	8's	9's	10's	Low Priority 11's
scheduled face to face meeting	58.75	18.75	10.00	6.25	3.75	1.25	0.00	0.00	0.00	1.25	0.00
scheduled phone call or voice conf	8.75	47.50	13.75	16.25	8.75	2.50	2.50	0.00	0.00	0.00	0.00
impromptu face to face meeting	10.00	7.50	21.25	20.00	12.50	10.00	6.25	5.00	7.50	0.00	0.00
impromptu phone call or voice conf	0.00	6.25	10.00	21.25	17.50	17.50	12.50	7.50	5.00	2.50	0.00
voice mail	3.75	8.75	8.75	12.50	10.00	27.50	17.50	3.75	3.75	2.50	1.25
pager	13.16	1.32	15.79	6.58	18.42	6.58	9.21	3.95	2.63	10.53	11.84
email	5.00	6.25	12.50	6.25	12.50	15.00	17.50	6.25	15.00	2.50	1.25
fax	0.00	0.00	1.25	5.00	8.75	7.50	15.00	31.25	11.25	13.75	6.25
fedex or priority mail	0.00	1.25	5.00	3.75	2.50	7.50	13.75	17.50	30.00	12.50	6.25
mail	0.00	1.25	1.25	0.00	3.75	2.50	3.75	13.75	18.75	33.75	21.25
electronicly published information	1.30	1.30	0.00	3.90	6.49	1.30	1.30	10.39	6.49	19.48	48.05

Percentage of respondants

Figure 2.7 Forced priority ranking of various communication media.

pager. They also have to learn to send low priority messages via another media. I have seen other teams successfully use pagers or e-mail as their highest priority media. What is best for your team will depend on the team's size, function, life span, available tools, and technical maturity. It is important to have the discussion up front and have a common understanding. I can't emphasize this point enough. I have never seen a distributed team consisting of three or more people where the members initially held the same view about the priority of communications modes, especially when the option of "face-to-face" was taken out of the picture. Once an agreement is in place, it can be reinforced with personal web pages and contract provisions.

When senders take responsibility for prioritizing messages, they transform the work life of the receivers. When information is prioritized, the receiver has control of when and how the information is processed. The receivers move from a reactive to a proactive mode. We examined successful distributed teams and we found that they had reduced the amount of information that was "pushed" at team members and increased the amount of information that team members "pulled." When I use the term "pushing information," I mean sending messages in a way that forces the receiver to react to them as soon as they are aware of their presence. Unprioritized messages are by definition "pushed information" because the receiver must immediately examine all messages to avoid missing high priority messages. If all the e-mail I receive is unprioritized, I'm stuck wading through 350 messages every morning just to find the one message where my boss is telling me I'm fired. Table 2.1 gives examples of "pushed" and "pulled" information. Immature, ineffective teams typically push between 90 and 100% of the information in their organization. High priority messages may need to be pushed, but 90% of the messages you send or receive are not high priority.

In Steven Covey's book *The 7 Habits of Highly Effective People* he uses a chart similar to Figure 2.8 to explain effective time management. He explains that your most productive time is spent working on tasks which are "important but not urgent."

Unfortunately American workers spent 70 - 80% of their time working on things which are "urgent but not important." When we push unprioritized information at our coworkers, we are forcing

Table 2.1
Pushed vs. Pulled Information

Pushed Information	Pulled Information
face-to-face meetings	electronic bulletin boards
pages	intranets
phone calls	Lotus notes
unprioritized e-mail	source control systems
unprioritized voice mail	document control systems

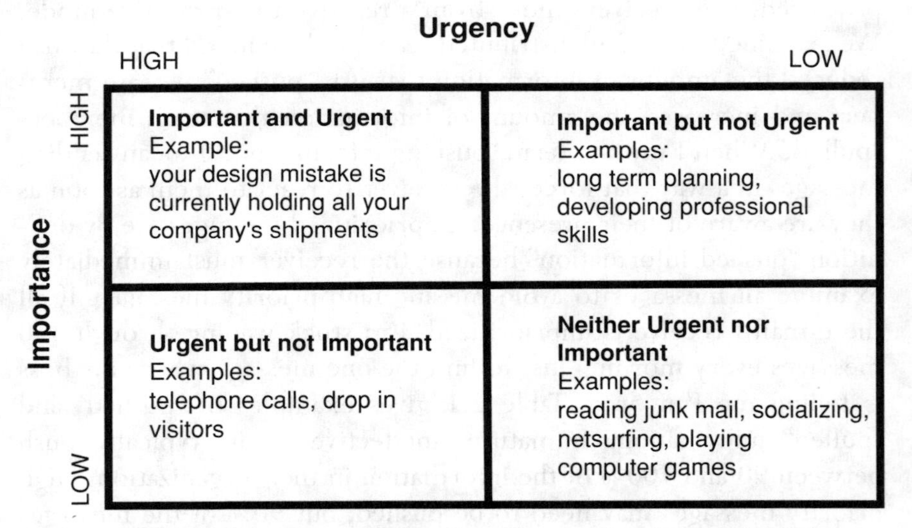

Figure 2.8 Time Management Matrix: From: Covey, *The 7 Habits of Highly Effective People,* (Simon & Schuster, 1989.)

them into the least productive quadrant of the chart. Studies show that it can take up to 20 minutes for a worker to regain focus after they have been interrupted. A team member would never dream of stealing something from a coworker, but they think nothing of walking into the cubical next door with a low priority question or picking up the phone and interrupting someone with an insignificant inquiry. Every time you do that, you are literally stealing time from your coworkers, time they could have used to complete the project earlier or go home and play with their kids. Poorly trained distributed teams live in a state of unprioritized broadcast information overload. They typically experience huge productivity hits because of the effect of information pushing on productivity and time management. By training team members to prioritize information and restructuring the way information flows between team members you can dramatically effect your team's performance.

Changing the way that information flows among team members should be a gradual process. One recipe for disaster is to suddenly decide that "everything is on the Web." Team members who are accustomed to having 99% of their information pushed at them won't adapt to a primarily pull strategy overnight. A 100% pull information strategy is almost never appropriate. When I have interviewed teams who have made a concerted effort optimize information flow, the ratio is still more like 40% push and 60% pull. Team members need to understand the business process before they can understand when to look for information. It is a good idea to put in place a transition plan. Be aware, however, that large teams may take several years to complete the transition.

The majority of this section has discussed changing behaviors when information is sent. There are some corresponding changes that need to take place on the receiver's side. Although you often hear team members complain about how they hate being interrupted, their behavior often doesn't always support their words. I've observed numerous team members drop whatever they are doing the moment they notice their e-mail indicator flashing or their voice mail light on. The little flashing icon seems to say "somebody wants me," "somebody needs me," "somebody loves me," "I'm important." You can almost see the seratonin levels in their brains go up as they bathe in the red light of their voice mail indicator. They choose to interrupt themselves and destroy their own productivity because they are completely addicted to their electronic connections. When I first observed this behavior, it reminded me of compulsive eating. Certain remote team members couldn't get anything done because they spent their whole day "snacking" on their e-mail and their voice mail. If they ate food the same way they input information they'd never be able to get in and out of their office. If you have prioritized communication and availability standards you only need to check your mail at specific intervals. Managers need to emphasize to team members that they will need to exercise some discipline in order to realize the productivity gains associated with the new communications infrastructure.

2.2 How are the four principles applied to various forms of electronic communication?

This section gives some specific examples on how to effectively use new technologies and suggestions for applying the key principles for distance communication. Your organization may be using technology that is more, or less technically advanced than the specific examples presented. The purpose of this section is to give you an idea of how principles could be applied. Technology is changing so rapidly that it is impossible to give a cookbook approach. The techniques presented probably need to be modified for your situation.

2.2.1 E-mail

In the early 1980s, Alan Cox did an extensive survey of communication patterns in American corporations. He asked both middle managers (Figure 2.9) and top executives (Figure 2.10) about the mix of formal and informal communication within their units. Sixty-six percent of the middle managers responded, "more or mostly informal," and sixty-nine percent of executives felt the communications within their unit were "more or mostly informal" [3]. If most of the business communications at your company take place informally (for example, hallway conversations instead of written status reports), you can see how managers might be afraid that team members who are out of the office might be out of the loop.

One technique for replacing lost informal communication is to create an "electronic water cooler." The three steps in this process are:

1. Create a symbolic name for the team.
2. Train team members to communicate with the team, not just individuals.
3. Use prioritization and filtering conventions to deal with "e-mail overload".

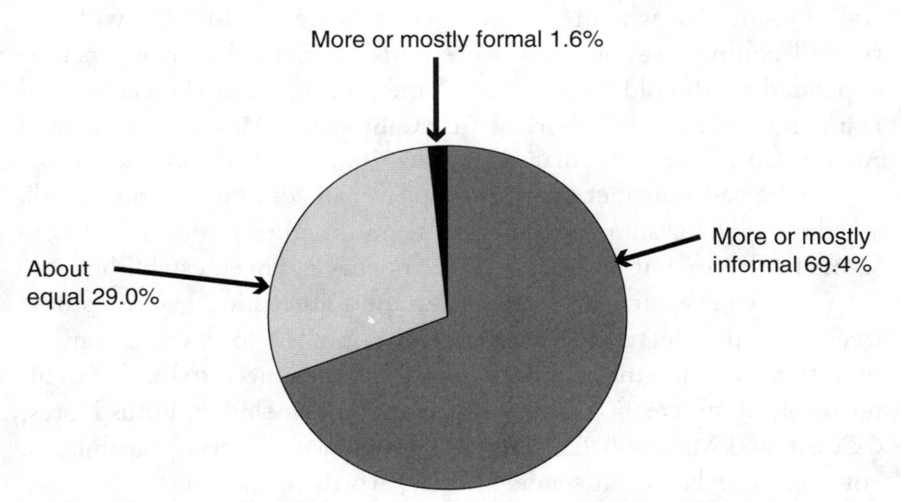

Figure 2.9 Survey of top executives' business communication patterns.

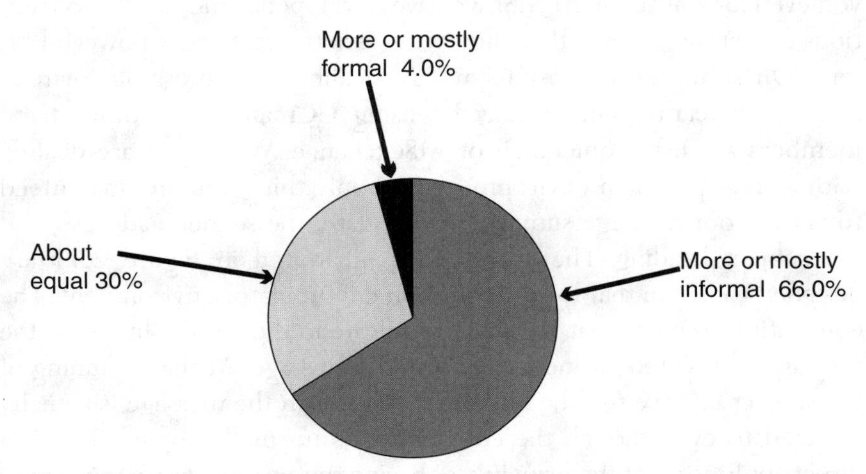

Figure 2.10 Survey of middle managers' business communication patterns.

Most e-mail systems allow you to create lists or symbolic names for the team. When I worked at Telebit, I was the project leader for a product code-named "LanBlazer." Our team had an e-mail symbolic name, "Lanb." Unless the messages were very private or confidential, they were directed to, or copied to the team. This practice made information equally available to all team members. We didn't have to depend on overhearing important status information over a cubicle wall. I found that when team members became comfortable with this type of communication, they were better informed than when they depended on the old "water cooler" method. On that project we had team members who worked in California, Massachusetts, and Amsterdam. That particular team was dominated by software engineers who had a distinct preference for e-mail communication. It took a little longer for some of the other team member to get used to the "e-mail culture," but within a year the norms had been established.

Once you've turned on the water and information begins to flow freely at your "electronic water cooler," you need to develop conventions to deal with e-mail overload. Team members need to be able to filter received information. Most e-mail systems (including Lotus Notes, CCmail, and Microsoft Exchange) have extensive filtering capabilities. You can filter based on sender's name, words in the subject line, the date, etc. Based on these filters, you can usually have messages stored in different folders, forwarded to your assistant, or even deleted before you even look at them. At MSI we have developed some simple conventions for sending e-mail that allow receivers to create very powerful filters. On some level, most teams are dealing with cross-platform e-mail. One team member may be using CCmail, while other team members are using quickmail or MSexchange. When you are dealing with a cross-platform environment, the only thing you are guaranteed to get in your message summary is the date, the sender's address, and the subject heading. The conventions illustrated in Figure 2.11 are designed to live in that lowest common denominator environment. The conventions require that the sender indicate in the subject line who the message is directed to and the priority of message. At the beginning of the subject line we use the initials of the person the message is actually directed to even though the entire team is normally copied. This is a direct application of the principle of having senders prioritize messages.

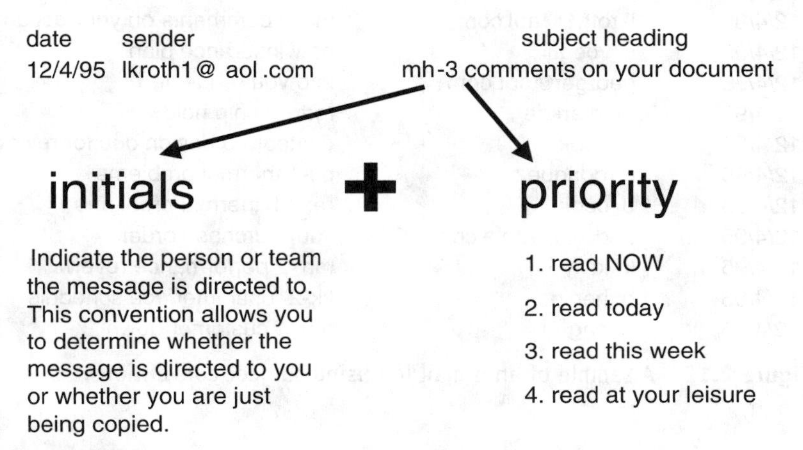

Figure 2.11 Subject convention headings.

The powerful part of this convention is that it allows the receiver to determine whether the message is being directed to him or whether he is just being copied. Even though some e-mail systems implement priorities, at the time of this book's writing, all of them assign the same priority to the receiver and people being copied, which isn't very helpful. Even if your e-mail system does not have filtering capabilities, using these conventions will enable you to visually filter your incoming information. Figure 2.12 below shows a sample of my e-mail from several years ago when I was using a system that did not support filtering. A quick visual scan of the subject lines shows me I need to read the messages from j_brown and r_long now. A quick scan of the sender column makes me aware of the e-mail I received from people outside of my team. In some cases the e-mail is directed to an entire team. Notice that in the seventh line of this example, the message is directed to "devteam." There is nothing magical about the four priorities or exact convention used in this example. The value is added when the senders prioritize the messages. Lockheed Martin's EIS group adopted a variation of these conventions. They include an R after the priority level to indicate that a reply is needed.

12/4/95	twila@risc.sps.mot.com	IEEE conference
12/4/95	j_brown	mh-1 today's meeting
12/4/95	lkroth1@aol.com	mh-3 comments on your document
12/4/95	c_voegtli	new insurance plan
12/4/95	georger@opt.com	info you requested
12/4/95	m_markee	re:td-2 ship hold
12/4/95	b_cook	devteam-3 design doc for review
12/4/95	l_rodriguez	bc-1 thermal problems
12/4/95	b_cook	re: lr-l thermal problems
12/4/95	arldole@venta.com	your purchase order
12/4/95	k_kelly	mh -2 performance reveiw
12/4/95	b_henig	kk-2 user interface schedule
12/4/95	r_long	mh -1 customer down

Figure 2.12 A sample of an e-mail log using subject conventions.

Now that we've slain the priority dragon, we can move on to talking about how to build context when we use e-mail as a communication media. The first type of context to consider is physical context. At one time or another, most of us have received e-mail messages that look completely garbled.

Generally, these messages have been transmitted "cross platform." Figure 2.13 is an example of a message I received that was supposed to have several columns of numbers.

Because the tabs on the sender's machine were in a different position from mine, it was almost impossible to determine how many columns there were supposed to be and what numbers were in what column. If you maintain the same "physical context," then the message on the receiver's screen should look like the message on the sender's screen. Here are some guidelines that will help senders create a message that looks the same on their screen as the receiver's screen. Try to use:

Carriage return instead of auto wrap
(Not every e-mail client has auto wrap. If you don't, insert a hard carriage return or your message will just dribble off the right hand side of some receiver's screens.)

Spaces instead of tabs

Courier instead of proportionally spaced fonts

```
Subj   :  Re:     mh  -1 stop shipments

Date:39  Sat, Nov 18, 1995 9:29 PM EDT

From:  74375.1667@           compuserve     .com

X-From: 74375.1667@           compuserve     .com (BOB
LEIGHTON)

To:  MCHaywood      @ aol  .com
(INTERNET:      MCHaywood      @ aol  .com)

cc:  dev  -team


Martha-

  Hereís the list of bugs we will either

  to be fixed           deferred
doc

 1124       1137       1162

    1145    1142             1111
1234

  1154  1136
```

Figure 2.13 E-mail sample.

Try to avoid:

Bold, italics, underlining, graphics, or colors

Including all of the previous replies in an extended e-mail dialogue will help build situational context for all the readers. Perhaps you have experienced the frustration of wondering what the issue was when you receive a forwarded e-mail that consists of only a reply.

Building social context via e-mail is more difficult. Videoconferencing is a much better media for that purpose. However, by writing a little more expressively we can give the receivers of our messages a better idea of what type of person we are. Because writing takes a little more effort than speaking, it is very tempting to fire off e-mail messages that have all the charm of traffic ticket. Six word messages like "When will you complete the design?" do not build social context or relationships. Spending some time projecting a little bit of your personality is worthwhile. You can use creative punctuation, smilies, or whatever is appropriate for your personality and corporate culture. I've had students in my classes claim they can't build social context in their e-mail because they are conservative people and their corporate culture is very conservative. Conservative and terse are not the same thing. The folks at Lockheed-Martin let me know that putting little smilies into e-mail was not necessarily a career building move in their corporate culture; however, they were able to come up with plenty of ideas for making e-mail more expressive.

2.2.2 Videoconferencing

Videoconferencing has been promoted as a technology that supports remote collaboration since AT&T introduced the Picturephone in the mid-1960s [4]. Most people generally attribute its lack of widespread adoption to high costs. Although that is partially true, the real reason lies in the fact that for the last 30 years it has been difficult to prove that videoconferencing provides any real benefit. Any sales person will tell you that regardless of how low an item's cost is, if there is no benefit, buyers don't bite. Until very recently, most of the studies conducted on video collaboration involved examining the effect of video on short, contrived problem solving interactions. All of the early studies (Williams, 1977, Oshsman and Chapanis, 1974, Conrath, 1977) came to similar conclusions:

> "…. there is no evidence in this study that the addition of a video channel" [to an existing audio channel] "has any significant effects on communication times or communication behavior" [5].

More recent studies suggest that the real benefit of video-conferencing is not in the individual meeting outcome but in the work process or social and relationship-building qualities of the medium [6]. Videoconferencing is really a tool for teambuilding. It makes no difference in the quality or speed of individual meetings. If you are going to meet with someone one time only, don't bother with a video conference. Use a teleconference instead. If you are going to have a long-distance conversation with someone you see in person on a weekly basis, don't bother with a videoconference. If you have a team with members in California, South Carolina, and Texas who don't meet face-to-face on a regular basis, buy some type of videoconferencing system and schedule a weekly meeting. Videoconferencing is a great way to apply the principle of regularly using synchronous communication as discussed in Section 2.1.3. I can give a personal testimonial about what a difference this practice makes with bi-coastal teams. I've used both methods and the teams that saw even a blurry version of each others' faces every week were much more cohesive, trusting, and (most importantly) responsive to one another.

Why is videoconferencing such a great tool for building relationships? Because it is the best and most natural tool for building social context. Although I briefly defined social context in Section 2.2.2, I'd like to go into more detail here to illustrate why videoconferencing technology works so well for building social context. When we have the appropriate social context, we have gone through what psychologists call a "social categorization and inference process" [7]. At the end of this process we decide what kind of interactions are reasonable and appropriate with our partner in communication.

> "Research has shown that we consider gender, ethnicity, age, personality types (such as aggressive, quiet, or manipulative), and social roles, both business related (engineer or executive) and non-business related (single woman, father, fellow tennis player). Much of the time these categorization processes occur rapidly and outside of our conscious awareness. They are often based on people's physical characteristics (facial features or clothing) and visible accouterments (briefcases or portable PCs), although other information, such as accent, conversational style, and job title, is also important." [8]

Many people are uncomfortable acknowledging that the process of social categorization and inference exists. Most companies have diversity training programs because certain inferences are not only immoral but also illegal. The fact is, research has shown a high degree of accuracy in people's judgments of the background knowledge, attitudes, and behaviors of members of specific social categories [9]. In this book, there is not time for a thorough treatment of the subjects of ethics or diversity. In most cases, we use social context to make our communication easier to understand and a more pleasurable experience for our communication partners. We are actually very good at using social context to improve our communication. Although a few people are unethical, I don't think that means we should throw the baby out with the bath water. For the last 40 years I've been developing skills that help me pick up signals that my communication partners are broadcasting to me about the way they want me to interact with them. If someone shows up for a face-to-face or videoconference meeting wearing a suit, a tie, and a Rolex watch, he is making a statement about how he wants me to interact with him. I assume he wants to talk to me more about business than technology and that he will be happier if my interactions with him are more formal (I will also wonder if he's trying to sell me something). If the same person shows up to a meeting wearing Birkenstock sandals, no socks, rumpled jeans, and a UNIX Expo T-shirt, I assume he is a software engineer and wants to have more technical discussions. I could be wrong, I'm usually not.

Videoconferencing can also be helpful in establishing shared physical context. It's great to be able to hold something up in front of the camera and have everyone know what you're talking about. However, unless team members understand how the technology affects shared physical context, there are a variety of traps team members can fall into. Videoconferencing systems can give you the impression that you know more about what is in your partner's visual field than you actually do. Some of the problems in establishing shared physical context include:

- Window size
- Window presence
- Window placement
- Confusion over which windows are public and which are private

- Camera angle
- Unseen participants

At MSI we have several types of videoconferencing units. We most commonly use the Intel Proshare and the Connectix units. Both of these units allow the receiver to resize the transmitter's window without permission or notice. In the case of the Proshare you can make someone's face about as small as quarter, as in Figure 2.14. Admittedly, there are some people you would just soon not look at or talk to, but if you are giving them the impression that you can see them when you actually can't, you are losing part of the communication. When people believe that you can see their gestures, they use them to transmit more information. Most people don't understand the size of the optimum visual field. Untrained or inexperienced video-conferencing users will tend to create a window that contains only the remote communicator's face. This is not limited to desktop videoconference systems. I've sat in far too many room-sized videoconferences where some frustrated Cecil B. DeMille has control of the camera and zooms in far too close on speakers. According to Fussel and Benimoff

Figure 2.14 Visual field is too small.

the optimal visual field is from the top of the head to the bottom of the elbow as illustrated in Figure 2.15.

I included Figures 2.14 and 2.15 because they recreate a personal experience. I was talking to a manager of a test department over a videoconference link and the camera operator had zoomed in closely on his face. I asked him, "Bob, how's the testing going?" and he responded, "Oh, just great." Because of my reduced visual field I didn't see his gesture of a down-turned thumb. He thought I had understood his sarcasm. Although I had detected a slight eye roll, without the gesture it wasn't enough to make me understand. A week and half later I was on the phone with him saying, "I thought you said it was going great."

Figures 2.16 and 2.17 show how window placement can violate the principle of maintaining shared physical context. It is so disappointing to be communicating and gesturing expressively, only to discover you've been under a spreadsheet for the last 20 minutes. Videoconference users need to either develop standards or actively inform one another about changes in the state of their screens.

Videoconference users also need to develop an awareness of the effect of camera angle. We use eye gaze and gesture to facilitate turn taking, confirm attention, manage pauses, and convey nonverbal signals. If there is a significant angle between the image of the person we are communicating with on the screen and the camera, our non-

Figure 2.15 Optimum visual field.

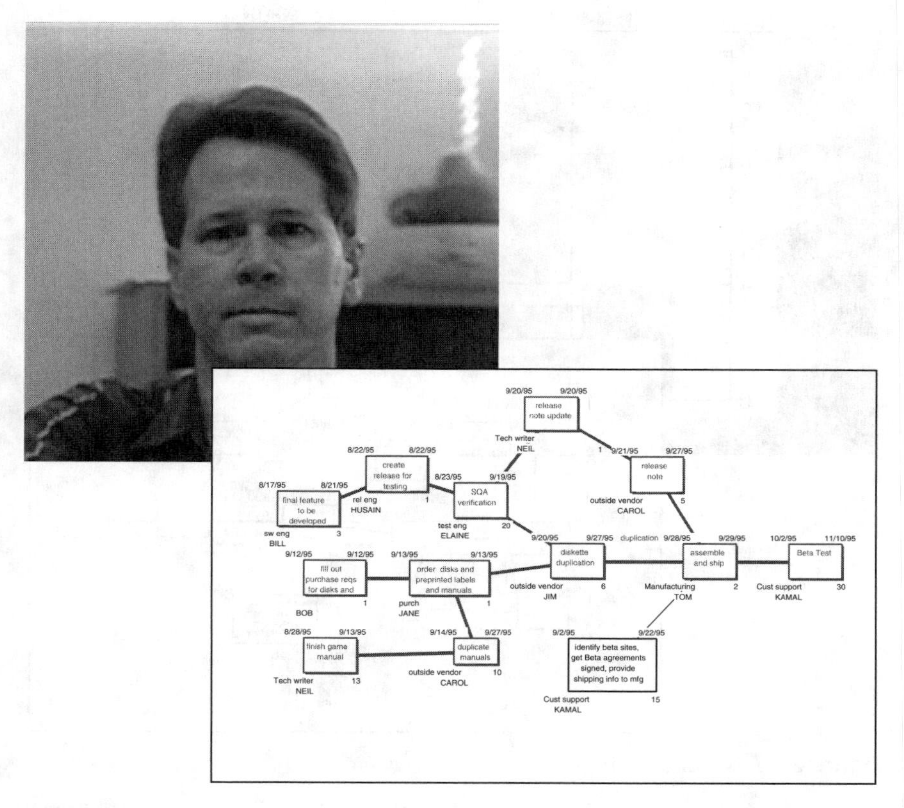

Figure 2.16 My Screen.

verbal signals become scrambled. It is not uncommon for older room-based video conference systems to have the cameras mounted up high on the wall while the screen that contains the images of the remote communicators is a table level. Even though the people in the local room are highly engaged and looking directly at the images of their remote team members, the remote team members receive an image that looks like the local team members are bored and sleepy (Figure 2.18). Looking directly into the eyes of the image of the screen does not necessarily provide the receiver with eye contact. You have to look into the camera. Most people to do not find this intuitive.

With some desktop units (particularly the inexpensive ones) it is easy to place the camera on top of the desk instead of the top of the

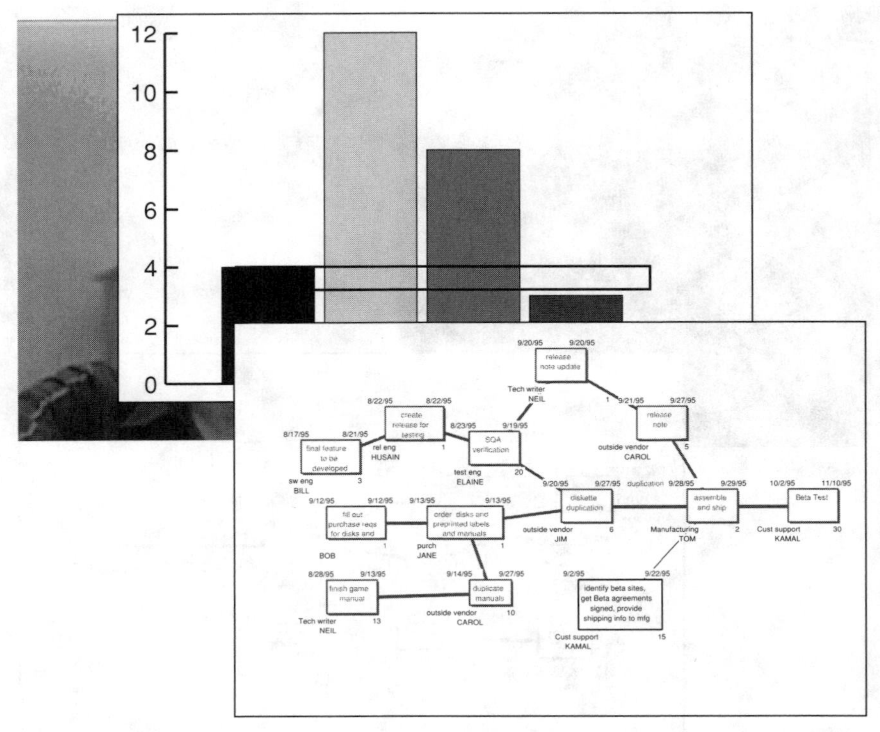

Figure 2.17 Your screen.

Figure 2.18 Camera angle too high.

Figure 2.19 Camera angle too low.

screen. This creates the unnatural view illustrated in Figure 2.19. It gives the receiver the impression that the transmitter is aloof. Researchers have determined that it is best to maintain a horizontal and vertical angle of less than eight degrees [10]. Many times it is possible to move the camera and get a very natural interaction. If it is not possible to move the camera, at least educate the team members. I have seen cases where people have decided their remote team members are disinterested and uncooperative just because they were being viewed from the wrong angle. Understanding the effect the media has on your messages is one of the keys to effective distance communication.

2.2.3 Teleconferencing

Voice or teleconferencing can be one of the most cost-effective means for meetings with remote team members. However, the media itself provides the participants with very few facilities for building physical, situational, and social context. The meeting members must be actively and conscientiously engaged in building context. The minimum standard of structure and preparation required for an effective voice conference is much greater than a face-to-face meeting. Co-located teams may be able to get away with casual or sloppy meeting management practices; distributed teams don't have that luxury.

Every meeting management class will tell you how important it is to send out an agenda. The Annenberg School of Communication study determined that, on average, 63% of business meetings in America have no written agenda [11]. If you have been attending meetings as I have for the past 20 years, this sad statistic is probably no surprise to you. An agenda is one of the primary tools for building situational context. Include a background section in your agenda summarizing events or situations leading up to the meeting. If you are the meeting leader, it is a good idea to contact the attendees individually before the meeting to see whether your background information is correct. At the beginning of the meeting it is a good practice to summarize the background and get agreement from the entire group that you have the same situational context.

Voice conferences leave a lot to be desired as tools for establishing social context. If occasional face-to-face or videoconference meetings are used in conjunction with regular voice conference meetings, social context may already be established for your team. Although I highly recommend a face-to-face kickoff meeting, I realize that for cost or logistics reasons, it is not always possible. In that case, the burden falls on the meeting leader to use other techniques to build social context. If the attendees have not met before, the leader should circulate an attendance list containing a short background of each attendee. The name and job title is really not enough. If it is a kickoff meeting for a team, putting together an e-mail package of all the team member's resumes is a good technique for helping to build social context. If a single new team member is joining a regular meeting, you should brief the new attendee privately on other team member's backgrounds and circulate a brief bio of the new team member with the agenda. I once attended a voice conference meeting where the project manager had hired an expert in a specific technology to attend a design review. Although the meeting leader introduced everyone by name, the team had no real idea what level of expertise the consultant could provide. Consequently, they directed almost no questions to him. Although the consultant wrote and submitted a report describing his recommendations, the team was denied the benefit of real-time interaction because the meeting leader did a poor job of building social context.

Establishing a shared physical context via a voice conference requires forethought and preparation. It is very important that you decide ahead of time which documents will be discussed in the meeting and make sure all of the attendees have them. It is a good idea to have a fax machine at each participating site so that spontaneously generated diagrams can be shared. The documents used in a voice conference must contain reference designators so that meeting participants can use words to point to specific pages, paragraphs, or diagrams in the documents. We don't think about how important it is to number pages and paragraphs because we are so accustomed to using our finger to point to what we're talking about. In our seminars we conduct a simulation where we pass out documents to two teams. The first team has documents that have no replacements for physical context (no page numbers, no paragraph numbers, and no designators for the diagrams). The second team has documents with good replacements for physical context (every page is numbered, every paragraph is numbered, and every diagram has a figure number). Then we conduct mock document reviews via voice conference. Team members try to describe their input to the document and communicate the changes that they want to make to other team members. The teams with the poorly annotated documents take up to three times longer to complete the exercise. Although working with poorly annotated documents is very inefficient and frustrating, many students tell me that at their companies, well annotated meeting documents are the exception, not the rule.

When team members are communicating via voice conference, lack of physical context can make it difficult to determine who is actually speaking and where conversion is being directed. When we are in the same room we use eye gaze and gesture to manage turn taking and indicate whom we are talking to. If I ask meeting attendees, "Can you see the screen?" while my gaze moves across the entire room and I make a sweeping gesture with my hand, the group knows I am directing the question to the entire meeting. If I say the same words while I look a particular person in the eye and point to him, the entire group knows I am directing the question to a specific person. For conversation to be appropriately directed in a voice conference all the participants must use verbal replacements for eye gaze and gesture to manage turn taking. We call this skill learning to "direct and self identify." The following example illustrates this technique.

> *Bob:* "The project will probably be about three months because we haven't been able to fill our last three requisitions for software engineers."
>
> *Jill:* "Bob, this is Jill. Have you thought about using some of the engineers in the Atlanta office? They have a project that is winding down."

In the example, Jill lets Bob know who is speaking and indicates that she is speaking directly to him, not the entire group. Although each one of us thinks our voice is very distinctive, the quality of the audio provided by voice conference technology, make it easy to confuse speakers.

2.2.4 Document conferencing

Although I love the idea of document conferencing and shared whiteboard software, at the time of this book's writing, the technology is not quite there. I have used most of the popular packages (NetMeeting, Proshare, and Cooltalk). Even over ISDN links, their slow performance makes them unusable. If your team's topology is such that you have very high bandwidth links between sites (384 bps or greater) and you can dedicate one of these links to a document conference, it may be worth your while. As bandwidth becomes cheaper, this medium will become a good way to replace physical context. I have seen it used almost always in conjunction with voice conferencing or videoconferencing.

2.3 How do you know what form of communication is most appropriate for the situation?

It is important to realize that you can mix a variety of electronic communication media to meet the needs of a specific situation. Using a voice conference in conjunction with an Internet link to web-based information might be a great way to make an interactive presentation to a distributed group. If you think about the types of context and the levels of interaction that are necessary, it's easier to put together the right tool set. Even if a meeting doesn't specifically require a very rich interaction, you may want to occasionally use a rich media to facilitate team building. The type of communication you choose for a specific situation depends not only on the task at hand but also the larger context of individual relationship building, team cohesiveness, time management, information flow, and workflow. Communication functions such as sending and receiving status can usually be accomplished via e-mail or electronically published media. Technical reviews usually require the interactivity of at least voice link and document-conferencing technology (or some thorough planning to deal with issue of physical context). Almost any media can be used to provide positive feedback; however, negative feedback needs to be delivered via a synchronous private media. Brainstorming meetings usually need to take place in a context rich environment. The questions to ask yourself when select one or more communication media are:

- What level of interaction is required?
- How will I build the required context?
- How will I ensure the communication is received, understood, and acted upon?
- How will I ensure the communication is appropriately prioritized?

References

[1] Fussel and Benimoff, "Social and Cognitive Processes in Interpersonal Communication: Implications for Advanced Telecommunications Technologies," *The Journal of Human Factors and Ergonomics,* June 1995, p. 229.

[2] Isaacs and Tang, "What Video Can and Cannot do for Collaboration," *Multimedia Systems,* 1994, pp. 63–73.

[3] Cox, Alan, *The Cox Report on the American Corporation,* Delacorte Press, NY, 1982.

[4] Isaacs and Tang, "Why Do Users Like Video? Studies of Multimedia-Supported Collaboration," *Computer Supported Cooperative Work,* 1993, pp. 163–196.

[5] Ochsman, Robert B. and Chapanis, Alphonse, "The Effects of 10 Communication Modes on the Behavior of Teams During Co-operative Problem Solving," *International Journal of Man-Machine Studies* v. 6, pp. 579–619.

[6] Isaacs and Tang, "Why Do Users Like Video? Studies of Multimedia-Supported Collaboration," *Computer Supported Cooperative Work,* 1993, pp. 163–196.

[7] Fussel and Krauss, "Coordination of Knowledge in Communication: Effects of Speakers' Assumptions About What Others Know," *Journal of Personality and Social Psychology,* V. 62, 1992, 378–391.

[8] Fussel and Benimoff, "Social and Cognitive Processes in Interpersonal Communication: Implications for Advanced Telecommunications Technologies," *The Journal of Human Factors and Ergonomics,* June 1995, p. 229.

[9] Fussel and Benimoff, "Social and Cognitive Processes in Interpersonal Communication: Implications for Advanced Telecommunications Technologies," *The Journal of Human Factors and Ergonomics,* June 1995, p. 229.

[10] Muhlbach, Bocker, and Prussog, "Telepresence in Videocommunications: A Study on Stereoscopy and Individual Eye Contact," *The Journal of Human Factors and Ergonomics,* June1995, 290–305.

[11] Creighton, James L. and Adams, James W., *Cyber Meeting,* Amacom, NY, 1998, p. 24.

3

Building a Team

WHAT MAKES A TEAM different from a group of individuals? When we ask managers, the answer we invariably receive is "common goals and values." If we continue the interview and ask, "If you could make sure all your distributed team members had common goals and values tomorrow, would your distributed teams function as well as your colocated teams?" Typically, after a few moments of reflection, the answer is "absolutely not," usually followed by a long discussion of reasons such as lack of trust, incompatible business processes, or problems with unreliable e-mail. If you think about it, competitors have the same goal but they are certainly not a team. The differences between competitors moving toward the same goal and team members moving toward a shared goal is their task interdependence, shared responsibility, shared process, relationships, and trust (Figure 3.1). I like Bob Kantor's perspective on teams.

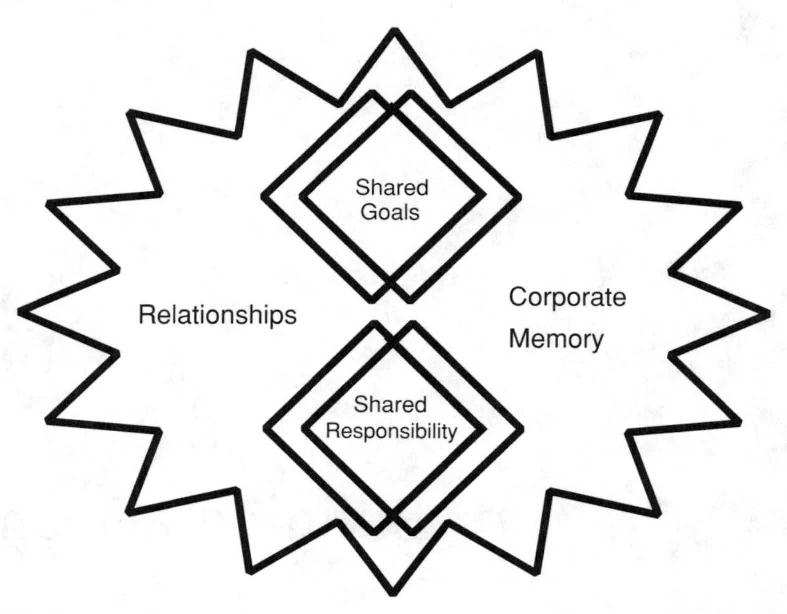

Figure 3.1 Elements that differentiate a team from a group of individuals.

"Teams are a higher form of organism than mere groups of individuals sharing information and tasks. They exist only after group members have made and demonstrated commitments to shared goals, visions, and values, and after they have articulated and reached consensus on roles responsibilities, and process. My experience with the care and feeding of high-performance teams suggests that the wiring together of talented professionals is necessary, but not sufficient." –Bob Kantor, Lotus Development Corp., in a letter to the editor in the May 15 *Fortune*.

3.1 What is the process for building a team?

Managers who have lived with virtual teams understand that team building requires managers not only to establish a vision but also create an infrastructure that supports communication and an efficient work process. The process of building a team involves:

1. Establishing a shared vision;

2. Creating an infrastructure (technology, policies, and processes which facilitate communication, workflow, relationship building, and corporate memory);
3. Selecting and assessing members;
4. Making the work experience rewarding and enjoyable for team members on a personal level.

At MSI we have developed two models that work in conjunction with one another to give managers a framework for building effective distributed teams. They are the "Alignment Model" and the "Maturity Model". The Alignment Model, described in Section 3.2, helps managers assess and select the team members who will perform the best within their current organization and infrastructure. The Maturity Model, described in Section 3.3, gives managers a framework for transitioning to the more advanced infrastructures that allow teams to achieve the highest levels of performance. Achieving the highest levels of performance requires that teams make adjustments to their goals, processes, tools, and skills. Implementing changes in the right order makes the difference between success and failure. One company that we worked with spent over 3 million dollars designing and rolling out a sophisticated Lotus Notes implementation. The software was intended to facilitate the work of their newly distributed teams. When we became involved, the company was in the process of abandoning the tool because team members were unable or unwilling to use it. The problem was not with the tool or the team members. Lotus Notes can be a great collaboration tool. The problem was with the way the management went about the implementation of their distributed teams. Installing sophisticated collaboration tools can be one of the steps for building a good infrastructure for a distributed team. Unfortunately for this company, it is not the first step. Distributed teams are multifaceted and complex in nature. You cannot start at the end or in the middle and work your way to the beginning (unless you want to waste a lot of time and money). Moving through the process in the right order is an exciting and enlightening experience. Doing the wrong things, or doing the right things in the wrong order, is painful and expensive. Section 3.3 discusses the Maturity Model in more detail. It will help you assess your organization now and determine the next steps you should take.

3.1.1 Key concepts and terms for using the models

In order to understand these models, it is important to define a few key terms and concepts. This section gives a brief explanation of the some of the concepts used in both models

Availability standards

In Section 2.1.1, I discussed how developing availability standards was one of the four key principles for effective distance communication. It is also a key concept for understanding the models. If you skipped Chapter 2, you may want to review Section 2.1.1 so that the models make sense.

Corporate memory

Corporate memory refers to whatever systems your team has in place to retain the knowledge to repeatably manufacture your product or perform your service. Historically, colocated teams have been able to get away with relatively informal systems for corporate memory. The corporate memory for some companies literally resides in the heads of certain long term employees. "Old Joe" has always been there. He's seen how everything has been done. If you have a question you can ask him. Unfortunately, the "Old Joe" method doesn't work for distributed teams. "Old Joe" didn't see the manufacturing process that your overseas partner used or the design process that your consultant created. Distributed team members need to feed into a formal system for corporate memory. Examples of systems for corporate memory include project repositories, document control, source control, and groupware. The systems and tools that would be appropriate for you would depend upon your industry.

When managers begin to manage distributed teams, they are most afraid of being unable to detect incompetent remote team members. While this is a risk, greater damage can be done by highly competent team members who don't feed into the team's system for corporate memory. To give you a specific example, about four years ago I was managing a project for a company that made large communication systems. Each system consisted of many printed circuit boards which plugged into a chassis. The company had outsourced the layout of some of their printed circuit boards to a small company that turned the layout work around very quickly. For the purposes of this example I'll call the small company "Layout, Inc." and my client "ABC, Co." For those of you not familiar with the process of making printed circuit boards, it goes roughly like this:

> A design engineer creates a schematic file which specifies all the components used and their interconnections. The file does not, however, show the exact physical placement of the components or the exact route of the interconnecting traces on the printed circuit board. That's what the layout engineer does.
>
> A layout engineer goes though a series of steps that result in a negative which shows the exact physical location of all the components and connections on the printed circuit board.
>
> The negative is used in the manufacturing process. Chemicals etch away all of the conductor on a printed circuit board which is not used for interconnecting traces and component pads.

One day the owner of Layout, Inc. decided to close the business. Shortly thereafter, ABC, Co. was forced to make a change to one of its printed circuit boards because a component vendor had obsoleted one of the parts. ABC, Co. soon discovered why Layout, Inc. had been so fast. Layout, Inc. had not been recompiling the schematics. The company had been making changes directly to the negatives (a process which is practically impossible to do accurately on a regular basis, but the owner of Layout, Inc. was one of those unique individuals). ABC, Co. had no idea whether the schematics that the design engineers were using actually matched the boards that were being manufactured. Recompiling the schematics now could introduce all kinds of problems. ABC, Co. was afraid to compile the schematics but had no one willing to try to make changes directly to the plot files or negative. Sadly, ABC Co. had essentially lost the recipe. Since they couldn't buy any more of the obsolete components, they couldn't manufacture their product using the old negatives. Recompiling from the new schematic would require a complete new layout cycle and verification test cycle. In this particular case, ABC, Co. decided it would be cheaper to obsolete the board than pay for a new layout and test cycle. This problem could have been avoided if ABC Co. had forced Layout, Inc. to submit the outputs of their intermediate work product into document control (the corporate memory system for hardware engineering departments). The hardware engineering manager would have immediately noticed the problem.

Process

Processes must be defined, documented, and placed in a corporate memory system before an organization can repeatably build a product or provide a service (Figure 3.2). The Software Engineering Institute and Carnegie Melon University have done significant work in developing methods for measuring process maturity for software organizations. Much of this information is available on the web at http://www.sei.org.

Pushing vs. pulling information

The amount of information each of us is expected to process each day is increasing. When information is presented to us in a manner that allows little control over when and how we process the information, that information is "pushed" at us. Examples of pushed information include phone calls, pages, voice-mail, and unprioritized or unfiltered e-mail. Examples of pulled information include electronic bulletin boards,

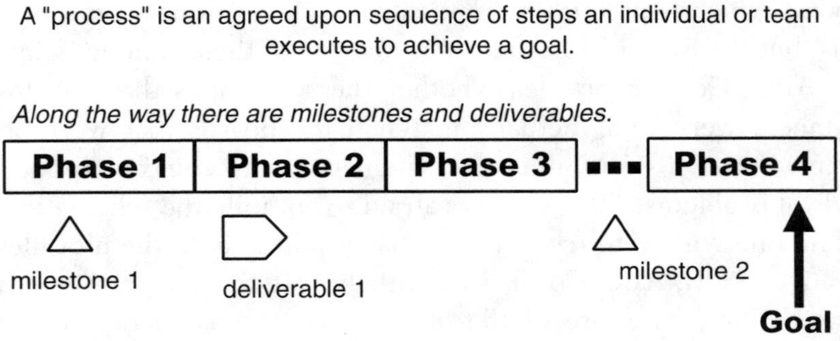

A "process" is an agreed upon sequence of steps an individual or team executes to achieve a goal.

Along the way there are milestones and deliverables.

Phase 1 **Phase 2** **Phase 3** ▪▪▪ **Phase 4**

milestone 1
deliverable 1
milestone 2
Goal

Figure 3.2 A process.

intranets, Lotus notes, source control systems, and document control systems. Several of the new means of electronic communication make it much easier for us to broadcast information to a large number of users. Poorly trained users of electronic communication enslave their coworkers in an overload of "pushed" electronic information.

3.2 How do I predict whether a person or organization will be effective as a member of my distributed team?

Many of the managers we have interviewed expressed the need for evaluation tools to help them select and assess team members. Only a small percentage of project managers have the luxury of selecting their entire team. Most of us who live in the real world inherit the majority, if not all, of our team members. Even if you don't have the option of immediately replacing certain team members, assessment tools are important because they allow you to predict problems and implement contingency plans. The Alignment Model is discussed in detail in the next section. It is designed to help you select and assess the team members who will allow you to obtain the best performance possible with your current organizational infrastructure. The case study in Section 3.2.1.5 shows how it helps to identify problems early.

3.2.1 Using the Alignment Model

The purpose of the Alignment Model (Figure 3.3) is to help managers get the best performance from their teams with the infrastructure they currently have in place. Section 3.3 discusses the Maturity Model, which gives guidelines for improving organizational infrastructure. In the Alignment Model, we envision each team member as puzzle piece consisting of four parts: goals, processes, tools, and skills. Figure 3.3 illustrates the fact that team members don't fit (or function as part of the team) unless they are aligned in all four areas.

Historically in colocated teams, managers have selected team members for their specific technical skills. Managers assumed team members would absorb the business process by osmosis and use the

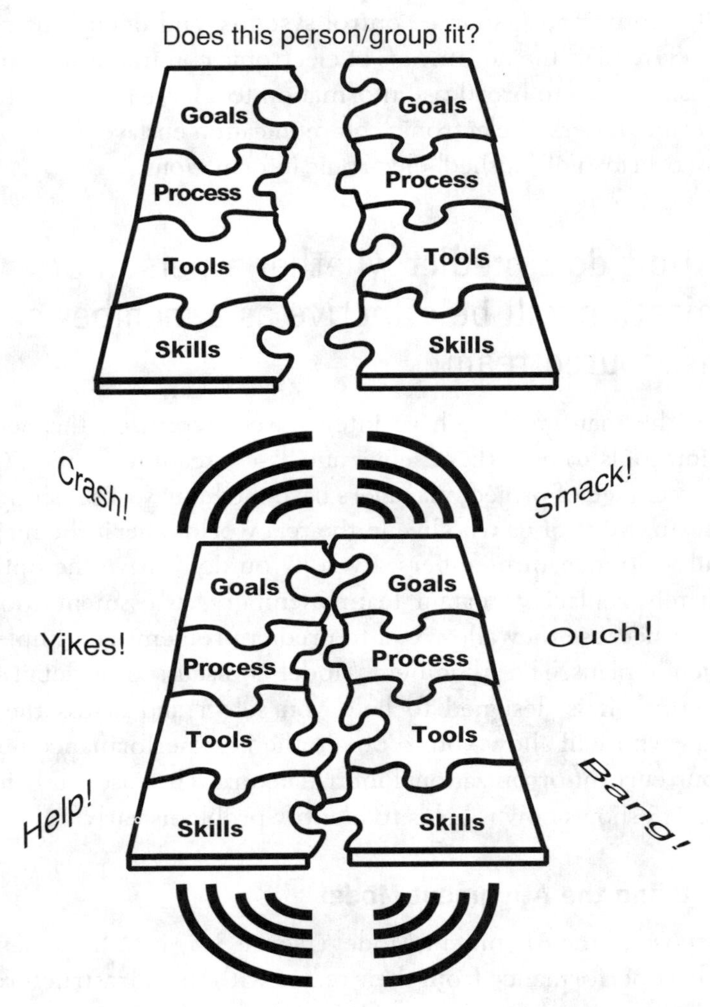

Figure 3.3 The alignment model.

tools provided at the central site. Unfortunately, osmosis doesn't work very well when people are thousands of miles apart. While technical proficiency is certainly important, it is sometimes actually more efficient to provide technical training to a team member who is already well aligned in the other three areas. Managers of distributed teams need to be much more proactive about defining and aligning goals, processes, and tools. When managers select and assess distributed team members, they are likely to find misalignments in several, if not all four, areas. If you are recruiting from outside your organization, it may be very difficult, if not impossible, to find the perfect fit. The idea is not to search for perfection, but to understand both the minimum criteria for alignment and how far the potential team member is from meeting these criteria. You can certainly hire a person or group which doesn't meet your minimum criteria, if you have a plan in place that will bring them into alignment. If you have no plan or infrastructure in place to teach a new remote team member your business process, expect a painful and expensive lesson on the value of aligning processes.

The assessment tools associated with the alignment model consist of a series of checklists. In Section 3.2 there are high level checklists for each element of the model (goals, processes, tools, skills). These checklists can be used for every type of team member. There are additional checklists in Appendix A that provide more detailed criteria for evaluating specific types of remote team members.

High-level checklist for assessing values and goals

Any project management course will tell you the first order of business is defining the project goals (Figure 3.4). It not surprising when you evaluate a deeply troubled project to find that the project manager is not only confused about the goals of the team members, he has done a poor job of defining and communicating the goals of the

Figure 3.4 Targeting goals.

project. It may sound obvious, but if you don't a written statement of project or team goals you are going to have a lot of difficulty evaluating team members for alignment.

In defining team goals and values, it is important to consider more than just dates and deliverables. Multi-site, multi-cultural teams need leadership in defining and communicating values. Believe me, the words "quality" and "on-time" mean different things to different people. Your team WILL NOT FUNCTION without a common definition.

The issue that we find companies have the most difficulty with is the definition of quality. One of my clients was the vice president of engineering at small company that created software products. The company created tools that helped technologists understand and install new technology. There was constant friction between the developers and the executive staff of the company. Every project was late and required the team members to work nights and weekends. The company experienced very high turnover among the developers. One of the fundamental problems we found was that the executives and the developers had very different visions of what was an acceptable level of product quality. The lead developers had extremely high standards and made it very clear they would not cooperate or put their name on any product shipment that had a known defect. The executives understood, but did not communicate that they were in an industry where their products were really only useful shortly after a new technology became widely available. Their products had a very short lifecycle and the market quickly saturated. Customers demanded the tools when they needed them and were willing to accept less-than-perfect software to have some help at the time of the installation. Lots of resentment developed between both the executives and the developers. What neither side seemed to realize was that it isn't the developers, project manager, vice president of engineering, or the director of quality assurance who sets the quality standard. It's the customer. If defect-free products were a requirement to be successful in the software market, Microsoft would be out of business. Although low-defect rate is an admirable value, in some markets, the customers are more interested in low price, fast time to market, or innovation. Nobody needs to feel good, bad, superior, or inferior about meeting market requirements. I'm sure

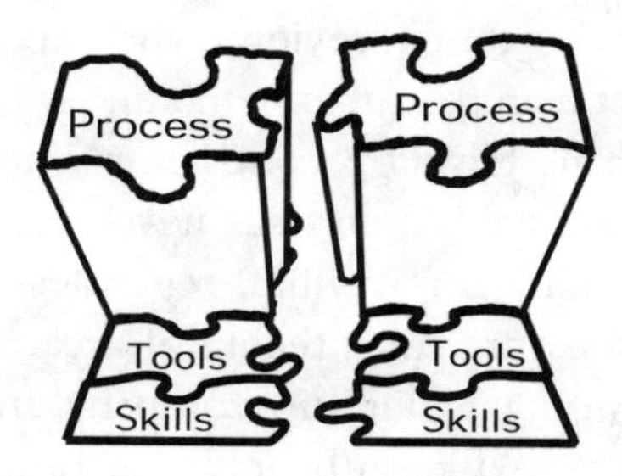

Figure 3.5 Targeting process.

the CEO of MacDonalds doesn't lose any sleep because his burgers aren't steaks. Some people want to buy a burger. Once the executives were able to honestly express the customers' values to the team, the team members who were uncomfortable with that standard left of their own accord.

The checklist below should help you evaluate the goals and values of distributed team members.

- What are their quality standards and how do they measure them?
- What is their reason for participating in the project?
- What is their attitude toward processes, tools, and skills?
- What are their competing priorities?
- What is their level of commitment to schedule and quality goals?
- What is their attitude toward decision making?
- What is their attitude toward communication?
- What are their cultural issues?

High-level checklist for assessing process

The area where we see the largest number of misalignments is in the work process (Figure 3.5). Project managers working in companies that have experienced mergers or acquisitions get hit hard with this issue. It is also crucial to pay attention to the work process when evaluating contractors and third-party development organizations. If your team is accustomed to a rigorous development process that includes writing specifications, conducting design reviews, and executing extensive testing, you need to select team members who understand that process and are comfortable with it. Hiring a contractor who has worked exclusively with start-up companies, and has never participated in a design review, is probably asking for trouble, regardless of how skilled that contractor may be in a particular technical area. The converse is also true. If you're working at a start-up company and you need to ship some kind of product next month or you'll be out of business, you don't want to work with a team member who requires a lot of structure. The checklist below will help you in assessing the processes of potential distributed team members.

- Do they understand what a process is?
- Have they documented their own process?
- What is their process maturity level?
- Do they understand your team's process?
- Do they have direct experience with a process similar to your team's process?
- How willing are they to adapt to the team's process?
- Do they have the tools available to feed into your team's systems for corporate memory?
- Do they have acceptable availability standards? Are they willing to commit to yours?

High-level checklist for assessing tools

How many times have you received an e-mail attachment that you can't read? Either you don't have the software application required to open it, or you have the wrong version. How many times do need to communicate with a team member who doesn't have a fax machine or an e-mail address? A little bit of up-front planning and communication about team standards for tools can greatly affect how quickly your project completes. Chapter 5 has a case study of a project where we analyzed the delays associated with misaligned tools. These delays added 25% the to length of the overall project.

Large companies that have experienced mergers and acquisitions have the most difficulty aligning tools. It is simply not economically feasible to throw out legacy hardware and software when companies as large as Lockheed and Martin Marietta merge. Alignment of tools does not necessarily mean that everyone has the same computer and software on their desktop, although that is one way to do it. It means that all the team members have equal ability to access shared resources and communicate information to the team (Figure 3.6). A Web-based infrastructure is an excellent solution for companies needing to align large legacy systems. Most of the commonly available word processors allow documents to be saved in HTML. Adobe's distiller and PDF format are excellent for making the outputs of platform-specific software readable by other team members. (By platform-specific software, I mean a special-purpose application, such as a computer aided design program that runs only on a Sun WorkStation.) The checklist below should help you align the tools of potential distributed team members.

- What tools (software applications and hardware platforms) will they have available for the work? If the tools are not identical, is there a common data-interchange format?
- What tools (such as e-mail, fax, voice-mail, file transfer protocol, or remote LAN access) will they have available for communication? Are the communication tools compatible?
- What tools will they have available for documentation? If the tools are not identical, is there a common data-interchange format?

High-level checklist for assessing skills

Most managers already do a pretty good job of assessing skills (Figure 3.7). It's the way they have selected their colocated team members

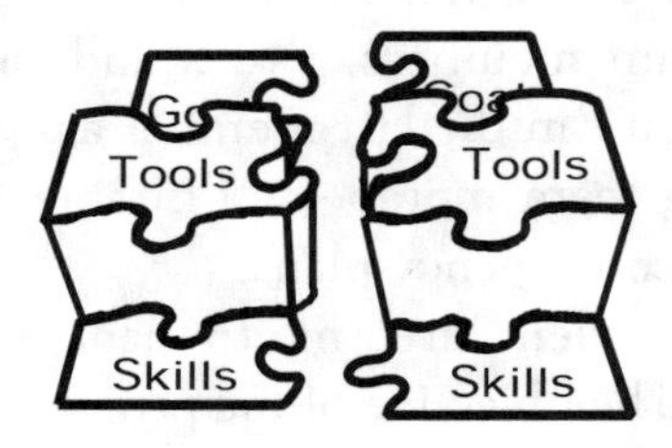

Figure 3.6 Targeting tools.

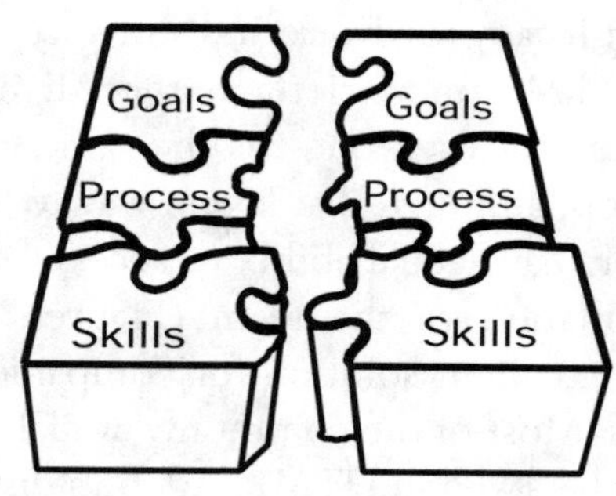

Figure 3.7 Targeting skills.

for years. For distributed team members, managers may want to examine a somewhat broader skill set. Distributed team members need to have better communication skills and also need to be able to do a better job of managing themselves.

- What is their level of technical skill and how can they demonstrate it?
- What is their track record for delivering quality work? delivering work that meets specifications? producing documentation?
- What is their level of management experience?
- What is their track record for delivering on time? delivering on budget? using the process?
- How strong are their communication skills?

Case study using the Alignment Model - TCT Technical Training Corp.

It is easy to see how the Alignment Model can be applied to selecting new team members. I'm often asked how useful an assessment is when you inherit team members. I've found the Alignment Model very useful, even when I'm the last member assigned to the team. By assessing the existing team members, I'm able to identify potential problems and create contingency plans.

About a year ago a client hired me to manage an internal information-systems project. The objective of the project was to customize and install a new software application for the sales department. The new software application needed not only to manage sales information, but also coordinate with existing software applications in the accounting and shipping departments. The project core team consisted of me, a systems analyst, the network administrator, an outside consultant, and representatives from accounting, sales, and shipping. The consultant, the systems analyst, and myself were all contract team members, who were only on-site occasionally. After assessing each of the team members it became clear to me that there would be serious problems with the outside consultant, Bob (not his real name). Bob had been hired because he had many years of experience with customizing the application TCT had selected. Bob's contract had been signed before I became involved with the project. There were no clear deliverables in the contract, no specific schedule and Bob had been paid 50% of his fee up front. This was the first red flag indicating that Bob's goals may not be aligned with the company or the team. After arranging the first team meeting, I found that Bob did not intend to participate in writing any type of functional or design specification, did not believe in detailed schedules, and would not be able to attend weekly team meetings.

At this point, another big red flag with the word "process" on it started waving in my mind. Since Bob's contract did not require that he write any specs, the other team members agreed to write the functional specs, providing that Bob would agree to review them for technical feasibility. We agreed that Bob could attend the meetings via voiceconference and that the team communication and document review would take place by e-mail and fax. At subsequent meetings we discussed team availability standards. Although Bob agreed to the standards, it soon became abundantly clear to the entire team that Bob had no intention of respecting them. When team members sent Bob faxes, e-mails, or phone messages he returned them only randomly and sporadically. Bob's fax machine was almost never operable. Bob traveled extensively and had no remote access to his e-mail. The other team members and I had no idea of when Bob might receive or reply to our messages. Needless to say, this is when the big red "tools" flag ran up the pole. It was very clear to me there was no chance for this project to succeed unless Bob made some major changes. After several discussions with Bob it was also clear to me that he had no intention of making these changes. Since I was a consultant project manager, I had no direct authority to hire or fire Bob. When I reported the situation to the vice president in charge, he was concerned but reluctant to let Bob go because the company had already paid Bob for 50% of the work. We discussed how the Alignment Model was usually very accurate in predicting potential problems. Luckily, it was relatively early in the project and I had set up a detailed project schedule. The vice president and I determined the point at which it would be impossible to recover from additional schedule slips. We agreed that if Bob missed all the milestones leading up to that point, he could be replaced. I began interviewing replacement consultants who were better aligned with the company and project. True to form, Bob missed every milestone. However, since we had replacements waiting in wings, we were able to successfully complete the project. Using the assessment process allowed us to manage risk, create a reasonable contingency plan, and recover from the selection error.

3.3 How do I measure the effectiveness of my team as a whole?

The Alignment Model is a tool for helping managers predict performance of team members in an existing team environment. How does a manager go about improving overall team performance? In our studies of distributed teams, we found that some teams were much more likely to meet their objectives than others. We decided to do some formal surveys and try to determine which characteristics were common to successful distributed teams and how those characteristics could be measured. These surveys resulted in the Maturity Model for Distributed Teams. This model gives managers a framework for assessing their team's maturity level and assistance in determining the next steps to improve their team's effectiveness. It works well in conjunction with the Alignment Model. Our experience showed that companies had a tendency to over-emphasize improvements in tools and under-emphasize improvements in processes and skills. The data showed that teams were more successful when they improved all four factors gradually. The following sections provide an overview of the model. Details of the model extend beyond the scope of this book.

3.3.1 The key characteristics of successful distributed teams

We found that there was a high degree of correlation between the following characteristics and a team's ability to meet it's objectives.

1. The existence of availability standards.
2. The reliability of electronic communication.
3. The existence of performance metrics.
4. Process definition, maturity, and alignment.
5. The existence of corporate memory systems.
6. The existence of written goals, objectives, and project specifications.
7. Managers and team members with a better-than-average ability to accurately estimate.
8. A lower-than-normal ratio of pushed to pulled information.
9. Team communication is prioritized by the sender.
10. Team member proficiency at distance communication.

3.3.2 Using the Maturity Model for Distributed Teams

The Maturity Model for Distributed Teams consists of four levels. Teams operating at each level have certain characteristics and key problem areas. One of the things we have found most valuable about the model is that it helps set the expectations of both managers and team members about how long it takes to transition from one level to the next. In large organizations (500 members or greater), it typically takes between nine months and a year to move up one level. It is not unusual for expectations to be completely unrealistic at all levels of an organization. Simply declaring yourself a virtual organization does not make it so. Team members, managers, and executives need to look at the implementation of a virtual team as a process that will take some time. For the purposes of this model "effectiveness" is defined as a team's record for meeting project or organizational objectives on time and on budget. Figure 3.8 illustrates the model.

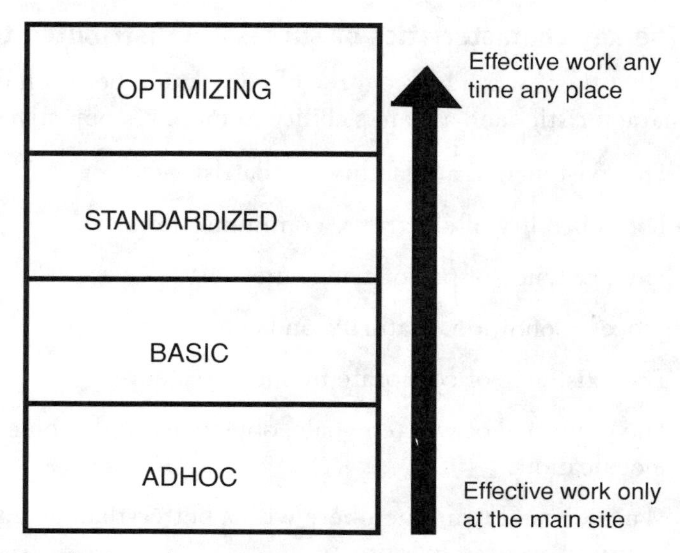

Figure 3.8 The Maturity Model for Distributed Teams.

Teams performing at the ADHOC level are typically out-performed by their colocated counterparts. Teams at the BASIC level typically achieve performance comparable to their colocated counterparts. Teams at the STANDARDIZED and OPTIMIZING levels consistently out-perform colocated teams. The following sections describe the characteristics and key problem areas associated with each level and give recommendations for moving to the next level. The key to the successful implementation of a distributed team is to move forward the goals, processes, tools, and skills of the team *together.* Implementing a sophisticated process without supporting tools and training in supporting skills is useless, just as implementing Optimizing level tools in a team with Adhoc level processes is a waste of money.

Teams at the Adhoc level

Teams at the Adhoc level consistently under-perform colocated teams. Some of the characteristics and key problem areas of teams at this level are:

Goals:

- Objectives are not stated or unclear.

Processes:

- There are no availability standards
- Business processes are misaligned and undefined.
- Systems for corporate memory are nonexistent or poor.
- Communication is primarily push.
- Management is by observation or "walking around."

Tools:

- Team member's access to electronic communication is nonexistent, unreliable or unequal.
- Tools are misaligned and incompatible.
- Performance metrics for team members are unreliable or nonexistent

Skills:

- Team members are not trained or are inexperienced with required modes of electronic communication.
- Communication is not prioritized.
- Team members have not been exposed to the principles of effective distance communication.
- Team members and managers have limited ability to accurately estimate resources and time.

The following recommendations are for moving to the next level.

Goals:

- Develop a written mission statement for the company or organization.
- Develop written, high-level project specifications and high-level objectives for team members.

Processes:

- Institute availability standards (see Chapter 2).

Tools:

- Stabilize the electronic communication in your organization. You must have reliable e-mail, voice-mail, fax, and file transfer facilities.
- Begin performance logging to facilitate the development of metrics and standards for team performance. (See Chapter 4).

Skills:

- Develop team member proficiency with the specific vendor implementations e-mail, voice-mail and fax.
- Develop a policy of having senders prioritize communications.
- Institute training on effective distance communication concepts (See Chapter 2).
- Institute training for managers and team members in estimating and scheduling (See Chapter 4).

Teams at the Basic level

Teams at the Basic level achieve a performance level similar to their colocated counterparts. Although they have begun to derive some of the benefits of a virtual organization, problems with infrastructure rob the team of time and efficiency. Some of the characteristics and key problem areas of teams at this level are listed below

Goals:

- Project specifications and team member objectives exist but are not sufficiently detailed.

Processes:

- Availability standards are in place.
- Business processes are misaligned.
- Communication is primarily push.

- The transition has begun from management by observation to management by objective.
- Corporate memory systems are inadequate or nonexistent.

Tools:

- Communication tools are aligned: electronic communication is reliable and all team members have access
- Application tools continue to need alignment.
- Performance metrics for team members are not yet stabilized, although performance histories are being maintained.

Skills:

- Team members have been trained in the specific implementations of e-mail, voice-mail, fax, and file transfer protocols.
- Team members have a limited understanding of distance communication concepts.
- The majority of communications are assigned a priority by the sender.
- Managers and team members have begun to improve estimating, scheduling, and objective writing.

The following recommendations are for moving to the next level.

Goals:

- Develop detailed project specifications and team member objectives.

Processes:

- Define, document, and align business processes.
- Institute processes for building corporate memory (coordinated with selected tools).
- Analyze information flow with the company. Target a subset of the information to transition from push to pull (coordinated with selected tools).

Tools:

- Select tools for implementing corporate memory systems and processes.
- Align application tools or select translation mechanisms.
- Select tools for transitioning information flow from push to pull.
- Develop performance metrics based on work history.

Teams at the Standardized level

At the Standardized level, the benefits derived from operating as a virtual organization outweigh the problems. Some of the characteristics and key problem areas of teams at this level are listed below

Goals:

- Organizational, project, and team members objectives are defined, documented, and aligned.

Processes:

- Business processes are defined and aligned.
- Processes and systems for building corporate memory are installed.
- Information flow is transitioning from push to pull.

Tools:

- Performance metrics, corporate memory systems, electronic-communication tools, and tools for information flow are in place and reliable.

Skills:

- Distance communication skills are well understood and practiced.

The following recommendations are for moving to the next level

Processes:

- Optimize business processes.

- Continue to analyze information flow.

Tools:

- Refine performance metrics.

Skills:

- Develop team member awareness of business processes (See Chapter 4).

Teams at the Optimizing level

Teams at the Optimizing level are characterized by the ability to have team members working any time, any place. New team members are easily integrated and released. At this level the team should continue to measure and optimize performance. The primary method for improving performance is the incorporation of new technology and business process re-engineering.

3.4 How do I make team members feel as if they belong to a team?

The majority of this chapter has been devoted to how to create a team that meets its objectives. Setting up an environment that facilitates the work process is only part of what managers need to do to build teams. Every manager knows that team members stay on teams they enjoy. Managers of distributed teams have to make the work experience rewarding and enjoyable for team members on a personal level.

3.4.1 Building team identity

The most important thing a manager can do to build team identity is to formally give the team a name. It is hard for team members to conceive of the team as a real entity if it doesn't have a location AND it doesn't have a name. If the only way your team members have to refer to their team is, "the guys who work for Bill," you're in trouble. Using the team name as an e-mail alias is usually very helpful for e-mail oriented cultures.

Using symbols like logos or pictures can also be helpful in creating team identity. Managers should encourage team members to use the symbols in all team communications. If it is culturally appropriate, you may want to give out T-shirts or mugs.

3.4.2 Facilitating relationships and building trust among remote team members

There are a variety of books on the market that identify the problem of "trust" as the most difficult issue associated with distributed teams. Maybe because we work almost exclusively with technical people (and let's face it, we're probably not the most sensitive group), we find that managers view trust as an issue, but not as their biggest challenge. In our practice we've found the most effective thing that managers can do to build trust among distributed team members is to establish availability standards. In Chapter 2 we discussed how to use personal web pages to allow team members to publish their commitments for responding to other team members. Several years ago I worked with a group of software engineers who had recently been to a team-building exercise. The facilitator had them navigate rope courses and talk about their feelings with their coworkers. I found it both interesting and amusing when one the engineers told me, "I don't care about this guy's inner feelings. I want to know when he's going to call me back." At that point, it occurred to me that trust has more to do with being able to predict a person's behavior than knowing their background or liking them personally.

Although it can be expensive, most managers of a distributed team will tell you it's a good idea to have some number of face-to-face meetings during a project. There's very little doubt that face-to-face meetings are one of the best and quickest way to build relationships. The most advantageous times to schedule face-to-face meetings are:

1. At the beginning of a project;
2. When there are significant personnel changes;
3. When there are complex or sensitive negotiations.

If there is a good electronic meeting schedule, we have found that teams should meet face-to-face at least once or twice a year. If there is no facility for regular voice or videoconferences, it is necessary to schedule more frequent face-to-face meetings.

Managers can also create other forms of electronic infrastructure that facilitate relationships among team members. I like to call them "electronic hangouts." An electronic bulletin board or an internal newsgroup for team members with common interests can serve as a replacement for the local hangouts where like-minded team members exchange ideas.

3.4.3 Making sure team members have fun

You remember fun—think back to before you were a project manager. Matt Weinstein wrote an excellent book called *Managing to Have Fun*. In it he discusses how people stay at jobs when they find the work experience rewarding and enjoyable on a personal level. He gives about 206 pages of creative suggestions for rewarding people. Not only should managers make it a priority to praise and reward team members, but they should set up peer reward systems so that team members reward one another. The tricky part for managers and members of distributed teams is knowing enough about remote coworkers to make the reward meaningful. I know one manager of a distributed team who decided to reward the performance of an outstanding ASIC designer, Bill, by having a plaque presented to him at a company meeting. Had Tom, the manager, taken the time to find out a little bit more about Bill, he would have known that standing up and receiving a plaque in front of a large group was more of a punishment than a reward for Bill. Anyone who knew Bill knew that what he enjoyed most in life was playing with video games and flight simulators. For the cost of the plaque, Tom could have given Bill a session at the local Magic Edge flight simulator. When we work in the same building with someone, we pick up all kinds of clues about their personal tastes by looking at the T-shirts they wear, the cars they drive, and the pictures they put up on the walls of their office. As managers of distributed teams we need to create a way for people to learn enough about one another to effectively reward each other.

One technique we have used effectively with clients is to include a link to some personal information on each team member's personal web page. Ask each team member to put together a few sentences about what they personally enjoy. Make sure they understand that the objective of the information is to have remote coworkers know enough about them to say thank you in a meaningful way. If you know someone loves jazz music, it's easy to have fun long distance by leaving a little clip of music on their voice-mail or sending them a tape when you really appreciate something. It is a good idea to give team members some guidelines about the content of personal web pages. (I live in California and some of things people enjoy can be pretty scary). It is best to avoid posting information about religion, politics, or other controversial issues. I often tell team members, "If you wouldn't hang it on the wall in your regular office, don't post it on the wall of your virtual office."

One of my partners has a client with a departmental web page. Each time you log on to the department web page the image of a different remote team member appears behind the picture of a reception desk. If you click on the team member's image you see information about that team member.

It is interesting that the smallest reward or positive interaction makes a big difference in people's behavior. There is a well-known psychology experiment nicknamed "The Good Samaritan Study." In the study, researchers observed individuals who made phone calls at a pay phone. They noticed almost everyone looked in the coin return after making a call to see if a coin was there. The researchers then randomly put dimes in the coin return slot. They hired a person to walk by the phone and drop an armful of books at the exact moment the study subjects were hanging up the phone. Interestingly enough, people who found money in the coin return were *four times more likely* to stop and help the woman pick up her books than the people who found no money in the coin return [1].

As managers of distributed teams, we need to be creative about the small rewards that make people feel good every day. What's appropriate in your environment may not work somewhere else. Our clients at Lockheed let us know pretty quickly that they don't use smiley faces in their e-mail, but they like it if you include a Dilbert cartoon in a fax or inter-office mail. There's nothing to stop you from having pizza delivered to the home or office of a remote team member on their birthday. Most managers know how to make people feel good if they want to, it is just a matter of making it a priority.

References

[1] Weinstein, M. *Managing to Have Fun* Simon and Schuster, 1997.

4

Remote Management Skills

WHEN TEAMS ARE geographically distributed, each team member will have to make adjustments, but the role of the manager probably requires the most change. This chapter will discuss some of the adaptations to the management process required for the successful operation of distributed teams. Some of the core management skills discussed in this section for managing at distance include:

- Managing by objective
- Mentoring and training remotely
- Developing a shared process

4.1 How do I know they are really working? How do I know they are working on the right things?

It is interesting to see how different management texts define the management process. Henri Fayol, who is credited with originating the process approach to management, believed the management functions were planning, organizing, commanding, coordinating, monitoring, and controlling. More recent texts tend to refer to the monitor, command, and control functions with less authoritative terms such as communicating, investigating, negotiating, evaluating, and risk management. Whatever you want to call it, both project managers and functional managers need to be able to determine the status of a task or project and make corrections to the project plan to compensate for problems. These days, it's a lot more hip to talk about "empowerment" than it is to talk about monitoring and controlling. All kinds of psychologists and management gurus would like to have you believe that if the team just "trusts each other and is empowered" you don't need to worry about monitoring the project. Experienced project managers know that detecting and correcting problems early is the key to bringing in a project on time. If the only effective method you have for monitoring is "management by walking around," being separated from team members can be a pretty frightening thought.

I find many managers confuse the functions of monitoring and controlling. They feel that if they collect adequate project status that they are, by definition, micro-managing the project and adopting an authoritarian management style. It is very possible to have open participative decision-making style while vigilantly monitoring performance. If you feel guilty monitoring your project, you should go lie down until that feeling goes away. Organizations learn and improve based on accurate performance information collected during projects. You may need to do some explaining, educating, and selling to your team members because they probably will associate monitoring with controlling. However, it is possible to get them to understand the difference and it is worth the effort. I know, because I've done it. In the book, *Designing Effective Work Groups*, Richard Walton and J. Hackman compare and contrast research conducted on work groups under "control" and "commitment" strategies [1]. A commitment strategy differs from a control strategy in the following ways:

- Team members participate in setting goals
- Performance and productivity information is shared with team members
- Team members participate in decision making and performance tracking

For either strategy to be effective, performance information must be collected. Although Walton and Hackman's writing clearly indicates they are proponents of the commitment strategy, they point out that they believe the best decision-making strategy for a group depends on the nature of the tasks and the nature of the group. Their research indicates control strategies are more likely to be effective with groups performing structured tasks. Locke and Schweiger (1979) published an extensive review of literature on employee participation in decision making, and concluded that participative decision making and goal setting had no consistent impact on performance when compared to assigned goals. They concluded that the critical factor was not participation in goal setting, but the acceptance of goals by work performers. My experience with engineers and scientists is that they seldom commit to goals or decisions in which they have had no voice. I use participative decision-making style with all my project teams. I wish I could say it was because I'm such a nice person and a highly evolved human being. The truth is, it is the style that gets the best results for me in the long term. If you are charismatic enough to get your team to accept assigned goals unquestioningly, my hat's off to you.

Whether you use an authoritative or participative style for decision making, there is no replacement for accurate project status and proactive response to variance. Make the decision yourself, or call a meeting, but get the decision made and the project back on track. Managers of successful distributed teams manage by objective and are able to create project plans, performance metrics, and reporting mechanisms that give them maximum visibility.

4.1.1 Managing by objective

Although you hear a lot a lip service paid to "managing by objective," not very many managers do a good job of putting the theory into practice. You can't blame them because there certainly aren't many good role models. As an individual contributor, I worked at several companies where my "MBOs" were reviewed only once a year. If my manager waited until the designated review time, it was far too late to implement any type of corrective action on the current project. In our survey of 514 high-technology managers, more than half of the managers normally allowed four weeks or more between scheduled deliverables. Monthly deliverables alone simply don't give a project manager adequate warning to make mid-course corrections. Project lifecycles in the high technology industries are usually a year or less. If you wait a month to find out about a problem, you can kiss your end date good-bye and start brushing up your resume. Since most good project managers do find out about problems earlier, you have to conclude that they are doing a lot of management by observation. Unfortunately, management by observation is simply not possible for distributed teams. The shift to managing by objective is not as easy as it might seem. Managing by objective is more than just defining deliverables. To successfully monitor remote workers you must:

1. Develop practical performance metrics.
2. Refine estimating skills of both managers and team members.
3. Increase visibility with frequent deliverables, prototyping and early integration.
4. Define project reporting mechanisms.

Developing practical performance metrics

The good news is that there is an incredible body of research on the subject of performance measurement. The bad news is that the science of performance measurement is a lot like meteorology. It is very beneficial but it leaves a lot to be desired in the areas of reliability and accuracy. There have been extensive studies conducted on performance indices, measurement scales, and rating methods. There are hundreds of thousands of pages of mind-numbing material that explores how the age, gender, and race of rators and ratees effect the stability, reliability, and accuracy of measurements. I've had the dubious pleasure of reading quite a bit of this material in the course of our research. The technical term for describing the stability, reliability, accuracy, and repeatability of performance metrics is "psychometrics." (I've never liked this term because it makes it sound like only crazy people measure things.) Most project managers don't have the time or inclination to become social scientists. At MSI we have created some simplified principles for developing performance metrics that will allow you to form useful metrics without drowning in detail. Even though the weatherman isn't correct one hundred percent of the time, the information he provides improves the quality of my life. In the same way, as members of project teams, we benefit from performance measurement even if it is far from perfect. In the succeeding sections, I will give a very brief summary of some of the principles of formal performance measurement and how we've adapted those principles into a practical approach for our clients. If you are interested in a more in-depth reference on formal methods I recommend, *The Measurement of Work Performance: Methods, Theory and Applications*, by Frank Landy and James Farr of Penn State University [2].

Performance management literature categorizes performance indices into two general types: behaviors and outcomes. Outcomes or outputs are the holy grail of performance management. They are typically easy to measure, but are spread out over time (in the project management world we call them deliverables). Behaviors are not as easy to measure as outcomes but they occur earlier and more frequently in a project. Many managers of colocated teams manage primarily by observation of behaviors. They note how enthusiastic a person is about the job, how long the person is working, whether the person asks questions of senior staff members, and how much time the person spends in the lab, among other things. I used to think that managers who managed primarily by behavior were lazy or simply didn't understand the project planning process. In my interviews, I found that many managers managed by behavior because their team members liked and encouraged it. Managing by behavior seems less authoritarian to both the manager and the team member. Some of the best managers in colocated environments have developed their own very subtle and complex behavioral metrics for determining whether progress is being made. Over a period of years they can get really good at it. They do it because it works. A survey of literature on performance management will tell you that research supports the fact that behaviors are predictive of outcomes.

When we work with clients to develop performance metrics for distributed team members, our first step is to ask managers to identify the behaviors they are currently using to predict positive outcomes for their colocated team members. If the managers are effective in their current colocated environment, it is usually not difficult for them to create a list. It is interesting that during this process many managers who rely heavily on observation of behaviors realize that the people they are managing are completely unaware of the criteria being used to assess them. They realize that keeping the criteria secret is a part of their strategy to obscure fact that monitoring is taking place. It's all part of the attempt to seem less authoritarian. I'll never forget the horrified expression of one director of engineering when he asked me, "You're not suggesting that I share this list with the people who report to me?". I had to tell him that I was suggesting exactly that. Here you have a list of behaviors that are likely to make team members successful. Because of your years of experience, you are confident enough that you base decisions on these criteria. Does that seem like a good thing to keep to yourself? It's a lot better to come clean with people about the fact that you are monitoring performance. Why should it be a secret? It is your job, after all. Explain the difference between monitoring and controlling, and enlist the team member's help in studying and improving the team performance. The best way to seem less authoritarian is to actually be less authoritarian. Adopt a participative decision-making style. Your knowledge about the positive effects of certain behaviors on outcomes is something that should be shared with team. When you are managing remote team members you really don't have an option for monitoring that's subtle. If you want to know whether remote team members are attending design reviews you're probably going to have to ask them.

If team members believe that you're monitoring because you don't trust them, you're headed for trouble. If they believe that you are monitoring because you are making sure they get their resources on time and working with them to optimize the team and individual work process, you won't have any trouble. Trusting that subordinates are competent is an area where many managers could use a little improvement. The Cox report showed that 41% of middle managers assume new subordinates are not competent [3].

The second part of our process is to decide if the behavior can be converted or treated as outcome. In some professions, this has been done for years. For example, sales people can be measured not only by the number of sales they close but also by how many sales calls they make. Making a sales call doesn't ensure a sale, but sales people who exhibit the behavior of calling on clients usually make more sales. Therefore, companies look at the behavior as an outcome, and provide some compensation based on the number of sales calls made as well as the number of sales closed. Landy and Farr indicate that the less control a worker has over the outcome the more important it is to use behavioral metrics.

Once you have decided what behaviors and outcomes need to be measured, it is necessary to determine how they will be measured and which rating method will be used. There are two general types of rating methods:

Judgmental - ranking or rating by supervisors, peers, or customers.

Nonjudgmental - measured by units produced, time elapsed, errors, scrap, absences, or similar objective means.

Although there is widespread dissatisfaction with judgmental performance measures, Landy and Trumbo determined that three-quarters of the performance measures in published research studies used judgmental criteria as the primary criteria variable [4]. Measurement of behaviors are most often judgmental. For example, most software managers believe that there is a relationship between having code reviewed and the outcome of a project. However, it is not easy to place a specific numeric value on how well code has been reviewed. Every engineer or technologist has been to a design review or code review where the discussion centered around grammatical errors and fruitless philosophical arguments. Evaluating design reviews by non-judgmental criteria (whether it happened or not) is not usually very helpful. Some rating scales for judgmental performance measures include:

Criterion-referenced scales, which are scales that evaluate performance of an individual or team without reference to other individuals or teams. Types of criterion-referenced scales are:

- Graphic-rating scales
- Behaviorally anchored rating scales (BARS)
- Mixed-standard scales
- Forced-choice rating scales

Norm-referenced scales, which are scales that compare the performance of an individual with that of another individual or group. Examples of norm-referenced scales are:

- Paired comparisons
- Ranking

Figure 4.1 below is an illustration of a graphic-rating scale. It is the simplest type of judgmental measurement, and has been around since 1922 when it was introduced by Donald Paterson. The relative ambiguity of response categories is addressed through what is termed the process of "anchoring." Anchors are an attempt to convey meaning about the various points on the rating scale. The scales B and C do a better job of anchoring than scale A.

At MSI, we recommend companies start with simple graphic rating scales for judgmental performance measurement. Over time you can improve the anchoring and you may be able to build a normative database which will help determine which behaviors are most highly correlated with positive outcomes. The development of most of the other types of rating scales (such as the BARS scale) is quite complex

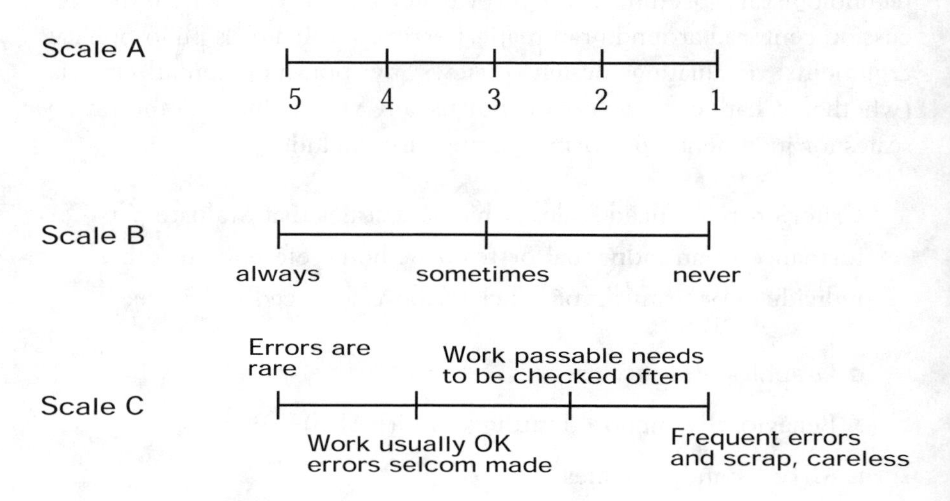

Figure 4.1 Graphic rating scales.

and demanding in terms of time and number of job experts required and they have not been proven to provide better measurements. To quote from Lundy and Farr's summary of research on judgmental performance measurement, "After more than 30 years of serious research, a generally accepted, efficient, and psychometrically sound alternative to the traditional graphic rating scale has not yet been developed."

The development and maintenance of performance metrics is an ongoing process. Many companies are hesitant to begin the process because they are afraid that the up-front investment will be too high. We have found you can develop something good enough to start with in a few days. The measurements will be a long way from perfect, but they will be a basis for you to review and improve performance over time.

Refining estimating skills

A key skill for managing by objective is becoming a good estimator. It's very hard to set realistic goals if you are a poor estimator of time and resources. I teach a class called "Project Management Skills for Engineers," which is designed, not for managers, but for individual technical contributors. One of the primary focuses of the class is developing accurate estimating skills. At the beginning of the section, I usually ask the students how many of them believe that they are good estimators. I usually see two or three hands raised out of a class of 25. I then ask how many students think their bosses are good estimators. The response is always underwhelming. I almost never see anyone raise their hand. Sadly enough, I find that a large number of technologists and managers believe that it is impossible to accurately estimate technical projects, particularly those working in the software business. Both our experience, and published research indicates that this is simply not true. PRTM's study of product development organizations showed that on average, only 28% of product development is really new. Most product teams are not completely new to their industry and technology. When I led product teams developing new modems or routers at Telebit, there were lots of pieces of the project we had done before. We knew how to design a case and power supply. We had laid out printed circuit boards before, and had submitted products to FCC testing. There were always new and risky aspects to the project, such as digital signal processing algorithms or technologies that we hadn't used before, but they didn't compose the majority of the project work. If you do a thorough job of estimating, even if your estimates are 100% in error on the parts of the project that are really new, your project shouldn't be more than 128% of your original estimate. Unfortunately, it is very commonplace in high technology industries for projects to complete two or three hundred percent over schedule and over budget. Unless a team is conducting basic research, there is no excuse for that type of performance.

If you cannot estimate now, distributing your project team is only going to make things worse. We always recommend that managers and team members take an estimating course before they start working in distributed teams. The subject of estimating deserves its own book, but we are asked about it so often that I think it is worth providing a few guidelines here.

Understand the objectives first Managers and team members need to resist the temptation to give off-the-top-of-their-head estimates. The road to project management hell is paved with people who give hallway estimates. There is always pressure to estimate projects that have one sentence verbal specifications. For some reason, project managers and technologists have trouble saying those three little words, "I don't know." Do your best to get the most information about the project that needs to be estimated. If you don't have a good definition of what the project or product is, your estimates are worthless.

Give a "date for date" One good strategy for dealing with the pressure to commit to specific end dates too early in a project, is to commit to a date when the project plan will be completed. You are essentially estimating in two steps. The first step is the plan. The second step is the project. Only when the plan is completed, do you really have an accurate estimate for the project. Personally, I have found this to be just about the best policy I ever implemented for myself or my teams. Once I started using this two-step process, I found that we started completing 18-month projects within 10 days of the date we committed to at the end of step one.

Break the project or process into component parts of one to two weeks When we are asked to review troubled projects, one of the first things we look at is the granularity of the work breakdown structure. If we see that the lowest level schedules have three or four-month task lengths, we know the project is going to be really late. Everyone is much more accurate at estimating a one-week task than a three-month task. The key feature of designing your work breakdown structure around one-week increments is that you can usually make a week up. You can work a couple of weekends or add extra resources. If you catch problems within a week you can almost always do something about it. If you wait three months, you're doomed.

Start with the familiar parts of the project As I previously mentioned only 28% of a project is really new. By starting with the familiar parts of the project, you can get a good head start on determining the magnitude of the schedule and budget.

Estimate the whole job If I ask a software engineer how long it will take to complete some software, the answer I'll usually get is how long it would take him to sit down and write the code. As a manager, when I say complete, I'm thinking, designed, coded, documented, integrated, and tested. When a software engineer hears complete, he or she is usually thinking coded. If you want an accurate estimate, be sure that you are specific about what the whole job is.

Allow time for vacation time, sick time and holidays It is a simple concept, but you'd be amazed how many projects are torpedoed by someone's honeymoon or appendix operation. On average, people are only available 80% of the time.

Estimate time to complete, not time on task When we work with technologists to help them improve their estimating skills, we often ask them to keep time logs for two to five days. This idea sometimes meets with resistance. However, after we assure them that they don't have to show it to their boss, they usually agree. At the end of the logging period we ask them to highlight the time they spent performing their primary work task. For example, if a test engineer's primary task for a week is to execute a specific test plan, he would highlight the time he spent actually executing the tests. Time spent on activities such as mentoring junior test engineers, answering calls from the customer support department, attending design reviews, interviewing potential coworkers, and attending company meetings would not be highlighted. The highlighted time is what we call "time on task." We have found the average time on task that among technologists in Silicon Valley is about 30%. For technologists to give accurate estimates they need to understand what percentage of the time they can spend on the task in their current work environment. Most high-technology industries are more schedule driven than cost driven. As a project manager I'm actually more interested in when it is going to be finished than how many hours it's going to take. At most commercial companies, an engineer's time is a research and development expense, whether he is performing overhead or project work. In some environments, however, particularly government projects or projects done on contract, you may need to track both time on task and time to complete.

Use "top down" and "bottom up" reality checks A "top down" reality check is the process of comparing your total project estimate to a similar project.

"The last time I wrote a file system it took three man-months."

A "bottom up" reality check is adding up the component parts of the project.

"The detailed work breakdown structure shows that the new file system will take 4.5 months."

If the estimates don't match, you should be able to identify a reason why.

Use historical data and keep records to improve future estimates The most reliable estimates come from a previous measurement of the same person doing the same thing in the same environment. However, even a measurement of a different person doing the same thing in a different environment provides a better guideline than nothing. It is a great idea to keep formal project and performance histories. But even if you don't have a formal repository containing that information right now, that doesn't mean there is no historical data available. I usually start with the person we think might perform the task. If they have no data, I try other people in the department or company. If no one in the company has performed a similar task, I try to get information from another company. Timesheets, project meeting minutes, or project schedules are the preferred source of data, but even a verbal account is better than nothing. Sometimes you actually have to hire a consultant to get some historical data about the tasks you are estimating. Believe me, if you are that far down the line, you are in a high-risk area of your project and need expert advise. NEVER pull an estimate out of the air. You are creating a monster which will stalk you throughout the life of the project. I've had to learn this particular lesson repeatedly in my career. It is tempting to just put in a few weeks for that little task I don't know much about and then hope I've got it covered. I have never gotten away with it. That little box on my pert chart hunts the project down like a bounty hunter and blows it to smithereens.

Increasing visibility

When I use the term "visibility," it encompasses not only the concept of monitoring but also the concept of verification. It is one thing for team members to sincerely believe they are making progress and report that progress to you. It is an entirely different matter to ensure that everyone's perception of reality is the same, and progress is actually being made. The best techniques for improving visibility include creating frequent deliverables, early integration, prototyping, and simulation.

Creating frequent deliverables The PMBOK (Project Management Body of Knowledge) defines a deliverable as:

> Any measurable, tangible, or verifiable outcome, result, or item that must be produced to complete a project or part of a project.

The word "must" in this definition implies that deliverables occur only when there is a hand-off of a project component between departments, coworkers, or customers. We have found that when you are working with distributed teams it is best to take a slightly broader view of deliverables. As a project manager, you may want to define deliverables the function of which is to provide visibility. The greater the number and frequency of deliverables in a project, the better the visibility. Care should be taken when defining deliverables this way. Give prime to how much it will cost to produce this deliverable, and how much it will cost to verify it. A good example of a deliverable that could be defined solely for the purpose of providing visibility is the outline of a design document. Downstream workgroups probably need the entire document to make progress; however, making the outline a deliverable helps sync up the team and verifies that progress is actually being made. The good thing about this example is that it doesn't cost much to create, and it doesn't cost much to verify.

Here is an example of what not to do. One software manager was absolutely paranoid that his remotely located programmers were not going to meet deadlines. To increase his confidence, he asked that the programmers e-mail him copies of the code as they completed writing it. He felt better when his mailbox was inundated with C++ source code. The few pieces of code he did have time to review seemed to be full of bugs. He quickly realized that verifying these deliverables was well beyond the scope of his abilities. The verification plan that he had in place was far too costly. It is much better to ask the engineers for unit-test results, or better yet, structure the project so that test engineers work concurrently with designers and provide independently verified unit-test results. If the deliverable can't be easily verified it is pretty useless.

We often use the following check list for defining deliverables:

Checklist for defining deliverables:

- Purpose: what is it for?
  Provide visibility;
  Hand-off information or product component between groups or coworkers;

- Quantification/specification: what is it?
- Recipient: who is going to receive it?
- Verification: how will it be verified?
- ROI,
  How much does it cost to produce?
  How much does it cost to verify?

Some techniques for creating frequent deliverables include:

- Asking for the outline of a planning document or specification;
- Asking for specific sections of a plan or document individually;
- Asking for prototypes;
- Using simulation techniques to prototype subsystems;
- Creating pilots or simulations of new processes;
- Holding design reviews.

Using early integration, concurrent testing, prototyping, and simulation Forcing team members to integrate project components early is an excellent technique for increasing visibility. Technologists are notorious for not wanting to let their work be seen until it is beyond reproach. This tendency is bad enough in a colocated environment, but it is poison for a distributed team. It is so easy for remote team members to work in a vacuum. If most of the integration takes place at the end of the project you can expect a lot of rework. One study showed that 48% of project work is rework [5]. As project manager, you have put the pieces of the puzzle together as early as possible, and find out about misunderstandings before too much effort is wasted. You will encounter resistance at every turn, but the payoff is great. It is especially important to plan early integration for virtual teams because it is not going to happen spontaneously.

Schedule testing and verification concurrently with design and development. If your company has an independent test department, the manager of that department should be your best friend. Get the test people involved early. It is a great way to increase your visibility and decrease your project lifecycle. If 25% or your project has been developed, 25% of your project should be in testing. This strategy requires that your work breakdown structure be by subsystem, because you have to plan for subsystems or subprojects to coalesce enough to enter independent tests throughout the life of the project. The "Waterfall Model" of product development has been pretty well discredited.

Prototypes and simulations are great tools for increasing project visibility. Simulations are particularly important when there are performance-related aspects to a project. Implementing a prototype or simulation early in the project is not without cost. However, it is extremely important to verify the load-related characteristics of a system early. If you cannot model it mathematically, you must build a prototype, or create or buy a simulator. You will easily recover the cost by catching problems in the design/implementation phase instead of the test phase. In some industries, it is easy to buy equipment that will help you create simulations. I worked in both the telecommunications and the data communications industries. It was easy to buy off-the-shelf equipment that would generate thousands of telephone calls or huge amounts of data traffic. Turning this type of equipment loose on a early prototype always yields previously unknown information. Sometimes your portion of the project doesn't involve hardware, or the target hardware won't be ready until late in the project. Software can be prototyped and simulated, too. You can write software that simulates undelivered hardware. You can write software that sends thousands of transaction requests to a prototype database manager or file system under investigation. Even if the database manager has only one object in it, you can still collect information about how quickly the software system can process a transaction under load.

Simulations are not just for product development or project managers. Functional managers in finance or customer service can create simulations of processes before they are fully rolled out. It's amazing how great that new payroll or call-escalation process works when you are only processing one check or handling one call.

4.2 How do I mentor and train remote workers?

We repeatedly hear that only very senior people can work away from the main office because junior people need to be at the main office in order to learn. I have also heard this line of reasoning used to justify confining senior people to the central site because they need to be present to provide guidance. The idea that people can only learn from face-to-face interaction is completely inaccurate. Face-to-face mentoring may be the only process for teaching currently in place at a company, but that can change very quickly. For technical skills, research has shown that the medium used for teaching has no effect on the student's learning outcomes [6]. You can use video tape, CD-ROMS, videoconferences, voiceconferences, web-based training, even a plain old instruction manual. The idea of distance education is not new. It is no secret that students prefer to be taught in a classroom or face-to-face setting, and there are certain management and communication skills that are best communicated face to face. However, the majority of skills required for a job can be effectively mentored from a distance.

We have found that it is best to look at the skills required for a remote worker's job systematically, then create a plan for team-member development. At MSI, we believe that mentoring needs to take place in five areas:

Technical Skills Mentoring in this area involves increasing proficiency in core business skills. Examples of this are teaching chemists about new chemical processes, or teaching programmers a new programming language. Technical skills are great candidates for independent, self-paced learning.

Work-process skills Mentoring in work process skills has to do with teaching team members how business is done on this team. There are both task-oriented and non-task-oriented work-process skills. An example of a task-oriented work-process skill for a software engineer is learning where and how to get copies of the current software release from the company server. An example of a non-task-oriented work-process skill is how to submit an expense report or a purchase order. Most "on the job" mentoring has to do with work process skills.

Communication skills In designing a plan for mentoring communication skills you need to determine what kinds of communication are key for your team. It is not just written and verbal communications anymore. E-mail skills, videoconference skills, and the types of distance-communication skills discussed in Chapter 2 are crucial to team success.

Leadership skills Remote workers need to understand how to motivate and influence other team members.

Management skills This area includes self-management skills (such as time management), as well as team management skills.

We put together the following checklist to help managers in one client's technical support department create team-member development plans.

Checklist for technical support analyst mentoring and development plan

Technical skills

Is technical-skills training available?

- Is there self-paced Web or CD-ROM based training for new skills?
- Is there a way for the manager to verify the training took place?
- Is there classroom training in the local area?
- Can training and reinforcement take place in short trips back to the main office?
- Is there a high-availability mentor in place?

Work-process skills.

Are the task oriented work processes documented?

- Exactly how do I receive calls?

- How do I put calls on hold?
- How do I create conferences?
- How do I document problems?
- How do I escalate calls I can't handle?
- How do I correct an erroneous entry I make in the computer system?
- How do I deal with an abusive and hostile caller?

Are the non-task oriented work processes documented?

- How do I submit a purchase order?
- How do I submit an expense report?
- How do I request equipment repairs?
- How do I report trouble with my telephone lines or data connections?

Communication skills

Are there assessment and training facilities available for:

- Written English skills?
- E-mail skills?
- Phone skills?
- Audioconference skills?
- Customer interface skills?
- Meeting management skills?
- Presentation skills?
- Negotiation skills?
- Leadership Skills?
- Building vision?
- Motivating?

- Influence management?
- Management skills?
- Staffing?
- Time management?
- Planning?
- Monitoring?
- Teambuilding?

4.3 How do I develop a shared process?

For a team to successfully work on a project together, they need the same understanding of how the work will be done, and in what order. Chapter 3 discussed the importance of aligning processes for virtual teams. If you are aligning the process of a team with members from many organizations, alignment won't take place overnight. There is a cost associated with aligning process and it requires some judgment to determine how closely aligned the processes need to be. In some cases, such as vendor-supplier relationships, extremely close alignment is not necessary. A good guideline for determining how much effort to spend aligning a team member's process is to determine whether other team members need to maintain or modify that person's or group's work. For example, if I hire a contractor to write a portion of the software for a communications project my team is working on, I want him in lock-step with my team's process. After the contractor is gone, my team will be responsible for maintaining and enhancing that code. I would include in the contract provisions that the contractor write design documents, attend code reviews, follow coding standards, and use our source-control system. On the other hand, if I am purchasing an entire sub-system from a hardware vendor and I have several alternative suppliers, I am less concerned about their internal processes.

A first step in creating a shared process is defining and documenting your own process. Creating a project lifecycle document is usually helpful. It doesn't have to be an encyclopedia. The most useful documents are about a couple of pages long. An old colleague of mine, Art Bailey, who is now the Vice President of Engineering at DSC Communications always said, "If you can't get a representation of your process on one or two pages, it is worthless." Figure 4.2 shows the lifecycle document used by one of our clients, NetFRAME Systems. NetFRAME builds high-reliability servers, therefore their products have both hardware and software components. Their product lifecycle consists of four major phases.

1. Concept;
2. Investigation;
3. Design and development;
4. Production.

The lifecycle fits easily on one legal-sized sheet and can be posted on every team member's wall. As you can see from the example, the design and development phases have several subphases: product design, development, integration, and SQA final. Managers from individual departments may want to create process documents with more specifics about the subphases.

Terms and meanings vary from company to company so it is a good idea to create a project glossary. For example, the term "alpha test" could mean the first time external customers test a product, the first time someone outside of engineering tests a product, or an internal test of a prototype. Figure 4.3 shows where "alpha test" has been at some of our client companies.

Merged and acquired companies usually have problems aligning processes. Much of the problem simply has to do with nomenclature. One of my partners, Virginia, says that the word with the most confused meaning is "done."

4.4 Influence management. How do I manage people who don't report to me?

I'm asked this question in every project management class I teach. Poorly implemented matrix management organizations are the source of untold personal and professional frustration. I've managed teams where all the team members reported directly to me and I've managed

	Concept	Investigation
Purpose of Phase	Exploring technical and market possibilities to meet customer needs	Formulate boundary conditions of project & proof of concept
Marketing	- Competitive analysis - High level product description	-Market research -Categorize prioritize Requirements - Preliminary business case - Product positioning statement - Marketing requirements document - Priority matrix - End of life analysis Product Corporate -related products Product transition strategy
Sales		Channel Plan
Engineering	Draft architecture	-Eng specification -System architecture -Project definition document with Resource requirements -Preliminary project schedule -Testing tools requirements identified -Performance plan
Manufacturing		-Equipment process review
Support		-Draft spare and support plan
Finance		-Initial product cost projection -Mix analysis -ROI -Product -Corporate-related products

Figure 4.2 NetFRAME product lifecycle.

teams where none of the team members reported to me. There's no question that it is easier on me when I write everyone's performance review and they know it. I used to believe the reason executive staffs set up matrix organizations was that they were just plain crazy and

Design and Development

Complete detailed product design conforming to specifications

Product Design	Development	Integration
-Warranty - terms and conditions -Marketing plan - Localization input -Preliminary forecast -Upgrade/FG packaging requirements	-Product introduction plan -Beta agreements in place -Define upgrade strategy -Beta plan	-Beta test -Final forecast
		VAR channel training plans
-First engineering BOM; components specified for proto builds -Design reviews -Preliminary test specification plans -EMC design input -Preliminary reliability goals -Equipment/HW acquistion plan	-Design spec completed -Performance tuning feedback -Early reliability testing -Test plan review -Evaluation tests (mechanical mfg., performance, interface & data integrity)	-Beta candidate software release -Final design review -Performance tuning complete -Design finished goods & upgrade packaging -Regression test software -HQA tests
- Design for test/mfg input - Material sourcing strategy - Identify new test process - Draft purchase part specification	-Manufacturing plans -Production equipment ordered -Qualify unique components -Forecast loaded MRP -Manufacturing BOM	-Manufacturing test available -Ramp plan and intro volume
	- Spare and support plan - Training plan	-Beta release notes and draft manuals -Beta spares available -Deliver internal training for beta -Product training plan

Figure 4.2 NetFRAME product lifecycle. *(Continued)*

didn't understand what it took to get a project done. As I slid down the slippery slope to middle age, I started reading performance comparisons of matrix, project, and functional organizations. I realized that there were other points of view. For example, my background is in electrical engineering. However, many of my project teams have included mechanical or component engineers. The times mechanical engineers have reported directly to me, I have not been the best mentor to them because I didn't know that much about the discipline.

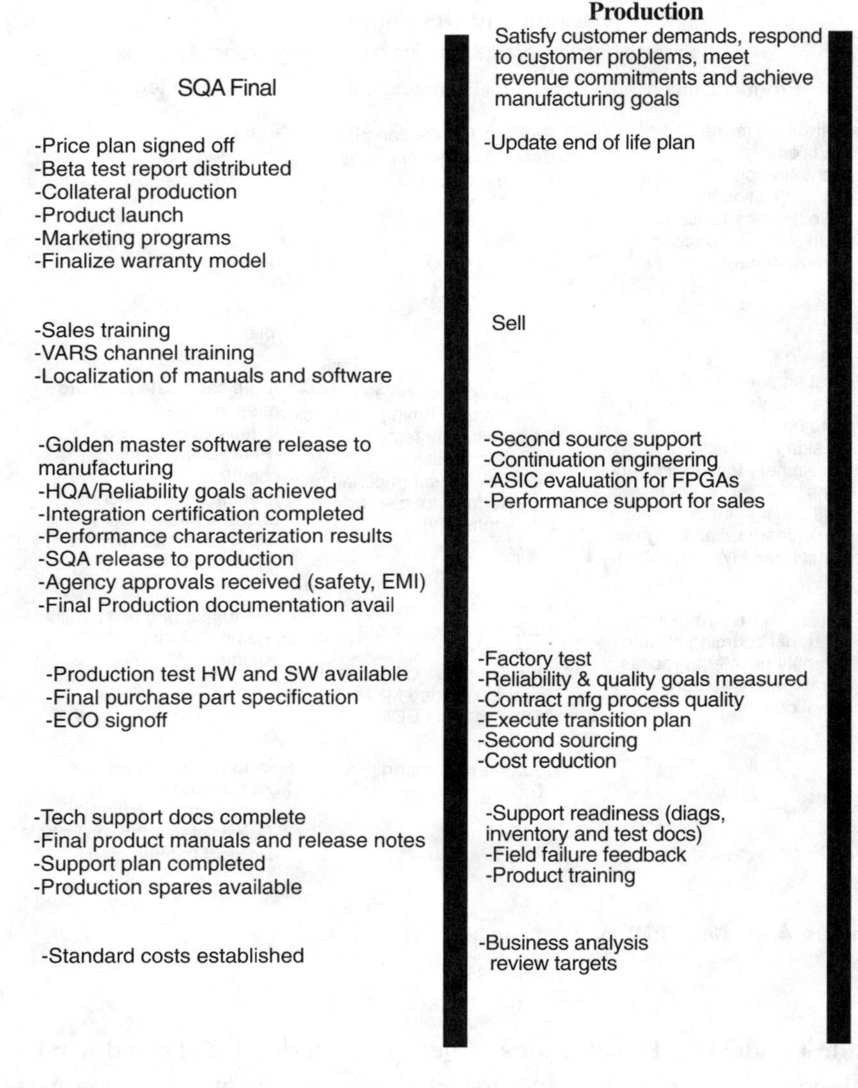

Figure 4.2 NetFRAME product lifecycle. *(Continued)*

Matrixed organizations provide a better environment for team member development and can provide better overall utilization of resources. However, there's a price to be paid for the benefits of a matrixed organization. It is a lot easier to command someone than to

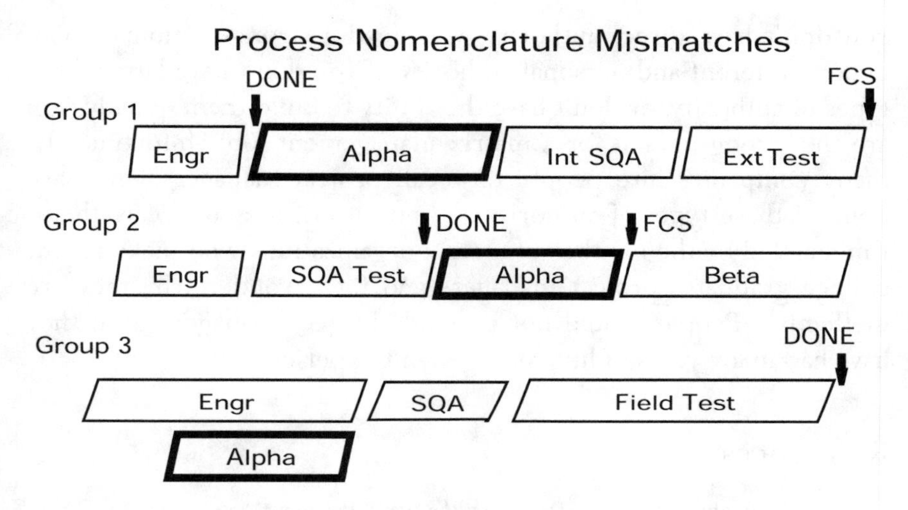

Figure 4.3 Process nomenclature mismatches.

convince someone. The extra effort required makes many project managers tired and cranky.

Assuming you have a fully or partially matrixed team, how do you get the authority to live up to your responsibilities? There are actually several types of authority you may have:

1. Direct authority: you have the authority to hire, fire, and reward and penalize team members.
2. Expert authority: you have more technical expertise than the team members.
3. Referent authority: you have excellent relationships with people in direct authority and are known to be a regular and respected reporting channel.
4. Personal authority: you have the ability to influence team members actions because of their personal relationships with you.

An executive at your company can give you direct authority in a day. The other types of authority take time to build. It takes time to build expertise in a subject and it takes time to build relationships. If you don't have direct authority, you need a strong combination of expert, referent and personal authority. If you don't have latter three types of authority, or don't have the ability to build them quickly, you are the wrong person for a matrix management job. Unfortunately, many companies hire people they call project managers who have none of these types of authority and no idea of how to create them. I'm startled by the number of matrix organizations who make recent college graduates project managers and then wonder why they are ineffective. People should not be made project managers until they have had many years of line management experience.

References

[1] Goodman and Associates, *Designing Effective Work Groups*, San Francisco, CA, Jossey-Bass, 1986, pp. 168-201.

[2] Landy and Farr, *The Measurement of Work Performance: Methods, Theory and Applications*, Academic Press, 1988.

[3] Cox, Alan, The Cox Report on the American Corporation, Delacorte Press, NY, 1982, p. 136.

[4] Cox, Alan, The Cox Report on the American Corporation, Delacorte Press, NY, 1982, p. 136.

[5] Smith and Reinertsen, *Developing Products in Half the Time*, Van Nostrand Reinhold, 1995.

[6] Willis, *Distance Education Strategies and Tools*, Englewood Cliffs, NJ, Educational Technology Publications, 1994.

Selected Bibliography

Bruns, *Performance Measurement, Evaluation, and Incentives*, Boston MA, Harvard Business School Press, 1992.

Heneman, *Merit Pay: Linking Pay Increases to Performance Ratings*, Addison-Wesley, 1992.

5

Organizing a Distributed Team

How closely tied does the structure of your organization have to be to the location of the team members? Your organization's maturity level (see Chapter 3 for a discussion of the maturity model) makes a difference in how you should organize your team. Organizations at low-maturity level do better with project-oriented organizations (Figure 5.1). In their case, entire projects or subprojects are assigned to different physical locations. It is important to assign a liaison at each site to coordinate communications between sites.

Organizations at a higher level of maturity can use functionally oriented, matrixed, or combination organizations (Figure 5.2). As shown, in this case, functional managers are colocated with team members at each location.

Organizations at the highest levels of maturity (Figure 5.3) can implement location-independent organizations. Team members can report directly to the project manager or be matrixed through functional managers who are also location-independent. Regardless of

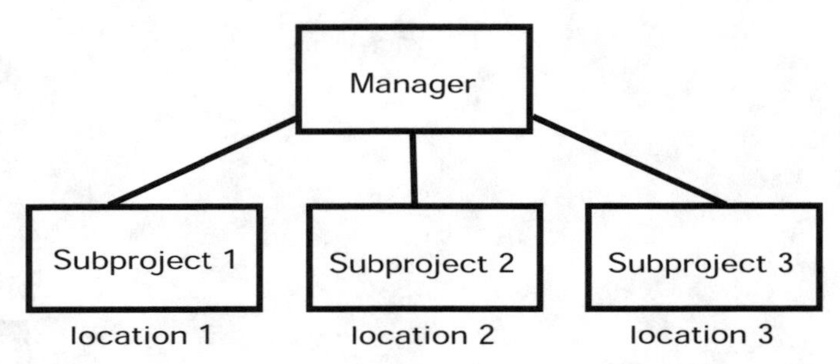

Figure 5.1 Project-oriented organizational structures are best for organizations at low maturity level.

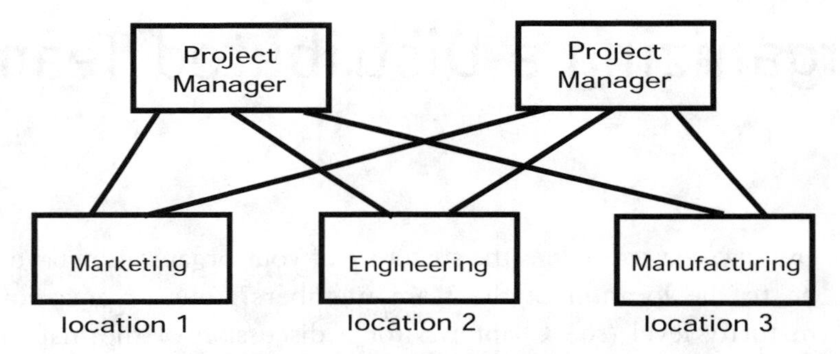

Figure 5.2 Functional or matrixed organizational structure can be used for organizations at a higher maturity level.

which organizational structure you implement, the succeeding sections will help you create a more efficient organization.

5.1 How do I optimally partition and locate work?

Making the right decision about who does what and where can make a big difference in your schedule and budget. This section will present some guidelines for organizing workflow and making work assignments.

Keep close collaborators in the same "time space." There are certain people on your project team who need to communicate with each other

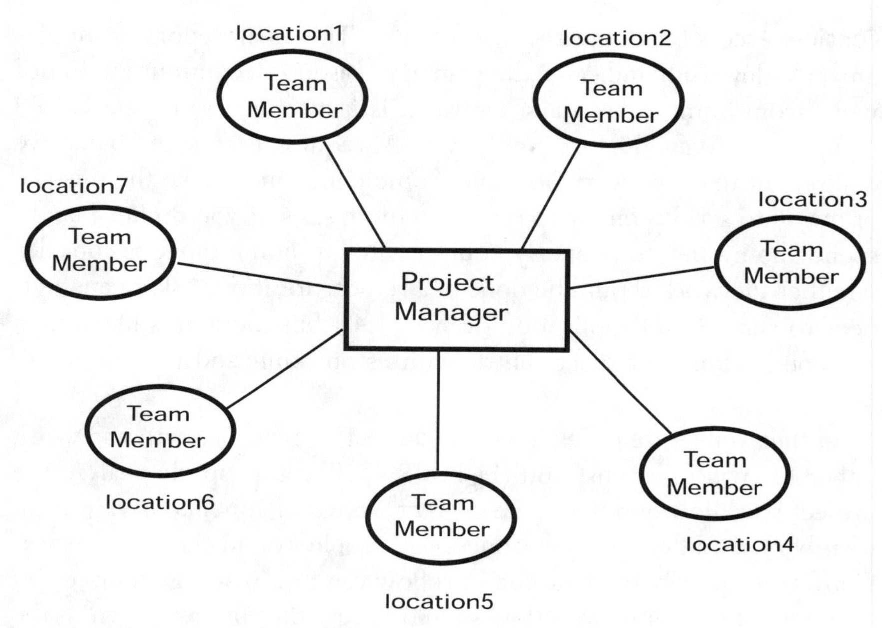

Figure 5.3 A location independent organization.

several times a day. At MSI, these are the people we term "close collaborators." An example of close collaborators would be an architect and a draftsman. We have found it is much more important for close collaborators to have the same working hours (be in the same time space) than it is for them to work in the same place. When people are working in the same time space, they are able to maintain a very high level of availability. Sometimes team members are close collaborators only for certain portions of the project. For example, a design engineer and a test engineer may need to collaborate closely when a new project enters the test phase. As a project manager, you need to identify the close collaborators at the beginning of the project, and determine how they will shift into a mutually acceptable time space. This doesn't mean that close collaborators must work exactly the same hours. However, in our practice we have found that a four to six hour-a-day crossover is the minimum acceptable time space. If the team members are not willing coordinate working hours you should do your best to assign the work differently.

Consider access to specialized equipment The final report from the Smart Valley study indicates the primary reason telecommuters do not work from home more days per week is lack of access to specialized equipment. Many high-technology jobs require access to expensive equipment that isn't very portable. Some companies make the mistake of trying to scrimp on equipment at remote sites. If you do not plan to spend the money to properly equip a satellite home office reconsider whether the work should be done at a remote location. Asking an engineer to run a CAD application over a 14.4 Kbps modem is like asking someone to build the space shuttle with a stone knife and a bear's tooth.

Make time zone differences work for you The turn of earth can work either for you or against you (Figure 5.4). If you properly analyze the project workflow and make the correct work assignments, having team members in different time zones can enable round-the-clock work. Conversely, poorly thought out workflow can lose working hours every single day. In general, workflow should follow the sun, as shown in the figure below. Best results come when workflow travels from east to west. Consider a hypothetical team where the software engineering department is in Boston and the testing engineering department is in San

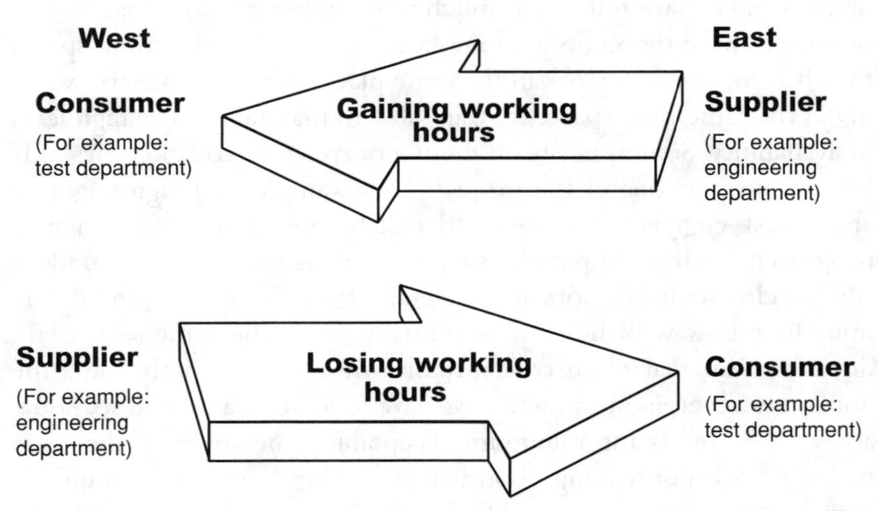

Figure 5.4 Let the workflow follow the sun.

Jose. If the software engineering department arrives in the morning and creates a release for testing, they will be ready to transfer it to the test department in San Jose at about noon, Boston time. For the engineers in California, it's only 9:00 a.m. The team has gained three working hours.

If the work were partitioned in the opposite way, with the software engineering department in San Jose and the test department in Boston, time would be lost rather than gain. If the San Jose engineering department sent the release east at noon, PST, it would be three in the afternoon in Boston. That's late in the day and doesn't leave any time flexibility. In the winter, by that time Boston people are often headed home early due to snow.

5.2 How do I eliminate the time delays caused by unnecessary transfers of physical deliverables?

Distributed project teams work best when the deliverables exchanged by team members are primarily information based. Electronic transfers such as e-mail, file transfer protocols or fax can make geographic distance unimportant. At MSI, we call work products which are not information based "physical deliverables." If I'm assembling a prototype of an electronic system and you send me the printed circuit board you have been designing and developing, that is a physical deliverable. It could take a day or two for the deliverable to travel between sites, even if we paid expediting charges. This is no big deal three or four times in a project. However, if the project work is partitioned in a way that makes repeated transfers of physical deliverables unavoidable, the timeline for the project can easily double. Examine the workflow and collaboration patterns of your project team. If you identify collaborators who must continually exchange physical deliverables, they should not be at separate sites. The case study below is an example of what happened when a company separated an electronic hardware engineer from the technicians building their prototypes.

Case study XYZ is a communications company with two primary locations in the United States, one on the east coast and one on the west coast. Historically, the west coast engineering department had kept a staff of technicians that worked closely with engineers to build, modify, and rework prototypes. As a cost-cutting measure, it was decided that prototypes could be built by the most senior technicians in the manufacturing department. As a result of a subsequent reorganization, manufacturing facilities were consolidated on the east coast. The technicians were now 3000 miles from the hardware engineers. Despite the numerous pleas of the project manger, the functional manager, and the hardware engineers themselves, the vice president of manufacturing refused let go of the turf. As the delays associated with this policy placed the project teams further and further behind, the hardware engineers began reworking prototypes themselves. This was hardly a cost-cutting measure, as the burdened cost for a senior engineer was three times the cost of technician. Because of the heroic efforts of several team members, the delay resulting from this improper partitioning of work was minimal. However, at the conclusion of the project, most of the technical leads and the project manager resigned.

5.3 What is a hidden physical deliverable?

Hidden physical deliverables are work products that should be transferred electronically but can't be because of mismatches in process or tools. How many times have you had to mail a piece of paper or a floppy disk because the file is too big to go through the mail system or the receiver doesn't have the correct version of the software application required to open the file? Each hidden physical deliverable causes a one or two day project delay. As individual incidents they make a minimal impact, but the accumulated lost time over the life of the project can be considerable. We have seen hidden physical deliverables add fifty percent to a project schedule. A common source of hidden physical deliverables is a poorly thought out authentication process. The case study below shows how a misaligned authentication process caused numerous delays for one company's project teams.

Case study After a merger, the ABC Company consolidated its purchasing functions at corporate headquarters. Emily (not her real name), a very senior purchasing agent, was made responsible for handling purchase requests from remote divisions. Emily had been at ABC for many years. At the time of the merger, she didn't have a computer on her desk. She didn't know how to use e-mail and wasn't particularly interested in learning about it. When someone at ABC headquarters needed to make a purchase in a hurry, they handed her a purchase order. She compared the approval signatures to the signature list she had in her desk drawer. If they matched, she phoned in the order. When managers from remote sites needed to purchase something quickly, they tried to fax Emily a purchase request. Unfortunately, the way the form was designed, the signature box was gray colored and it made the signature difficult to read when it was faxed. Emily had been trained never to make a purchase without appropriate authorization and consequently asked the remote managers to either send purchase orders via Federal Express or interoffice mail. This process created a "hidden physical deliverable" and added a two to five day delay in every purchase, greatly frustrating the remote managers and team members.

These types of delays are completely unnecessary. By analyzing your project plan and aligning the processes and tools of the team members, you can completely eliminate hidden physical deliverables. Had there been even a cursory examination of the purchasing process at ABC Company, it would have been obvious that is was inappropriate for a distributed team. There are plenty of solutions to the authentication problem (such as electronic signatures) which work perfectly with a distributed team.

The most systematic procedure for identifying and eliminating hidden physical deliverables is to closely examine the pert chart for your project. Note the points where there are transfers of information-based work product. Compare the processes and tools of the team members involved. You may have to interview team members to find out what tools they have available and how their processes work. This may seem like a tedious process, but you will recover the time tenfold during the project implementation. One of our partners, Virginia Lacker, works primarily with international software development teams. It is her policy to dry run every new type of information transfer and process she can identify before the project moves into implementation phase. She is quick to point out that just because you *should* be able to transfer a file between San Francisco and Calcutta over regular phone lines and a compatible modem doesn't mean you can. Also, a verbal description of remote equipment or software is not always enough to guarantee alignment of tools. In one project she managed for Harris Corporation, electronically transferring files in and out of Russia was so difficult and the mail was so slow, that engineers regularly flew from the United States to Russia to deliver hard disk drives.

6

Networking Technology

THE INFORMATION presented in this chapter is designed to help managers who are unfamiliar with remote networking technology become informed consumers of technology. If you are a networking professional or IT manager you may want to skip this chapter. This chapter will not prepare you to design, install, or debug a network. It will help you develop the vocabulary necessary to express your team's requirements to a networking professional. Networking technology is changing on a daily basis. Because it is such a crucial part of the infrastructure required by distributed teams, managers need to understand the nature and magnitude of the potential risks introduced by this new technology. New infrastructure always presents a project risk. Successful managers of distributed teams make contingency planning for infrastructure problems a part of the project plan. Many of the managers that we interviewed were unpleasantly surprised by one the following networking technology challenges:

- Equipment incompatibility
- Equipment unavailability
- Software incompatibility
- Infrastructure and equipment complexity
- User learning curves
- The magnitude of infrastructure maintenance and support
- The cost of equipment and communication facilities
- The difficulty of securing a remote access network

Although, it is tempting to think "Our company has an IT manager so it's not my problem," your IT manager may not fully understand your requirements and expectations. He doesn't own your project goals and deadlines. The virtual manager who isn't network literate will be consistently disappointed.

6.1 What are the basics of networking and remote access?

Networks are groups of intelligent nodes that communicate to share resources and information. The nodes include, but aren't limited to, personal computers, printers, mainframes, file servers, routers, telephone switches, and videoconference equipment. The nodes in a network can be permanently connected or they may connect with one another on demand. In order for the nodes on the network to communicate, there must some means of physical connectivity (a scheme for addressing), and they must use common communication protocols. Entities on a network need to be compatible on many levels to communicate. Figure 6.1 illustrates the seven layers recognized by the international standards organization.

The names of the layers and the functions performed at the layers are less important than the concept that several levels of compatibility are required. (For further reading I suggest *Client Server Computing* by Elbert.) If you and I were to use one of the old cup and string apparatuses we used to simulate telephones as children, we would have connectivity at the physical level. If I were speak into the cup and say, "Je suis tres heureuse de faire votre connaissance," you will hear me, but you may or may not understand me. Nodes on a network are the same way. Just because two devices have ethernet connections doesn't mean they can communicate. As project managers, we have to realize the level of risk involved with integrating a new piece of equipment or software into our network infrastructure. The opportunities for incompatibility are infinite. It is best to arrange a demonstration on your network before you purchase or begin to depend on a critical tool or application. I always ask as many questions about compatibility as possible, not because I don't trust the IT manager, but because I know it is a high risk area.

6.1.1 LANs

The acronym LAN stands for Local Area Network. All of the nodes in a local area network are typically permanently connected physically by some cabling medium. LANs were originally designed with the idea that all the nodes would be in same building on the same campus. For example, the ethernet standard (10Base2) has a maximum segment length of 200 meters [1]. In this section, I will use the term "interconnect technologies" to refer to the lower two layers of the communication protocol model (Figure 6.1). These two layers define

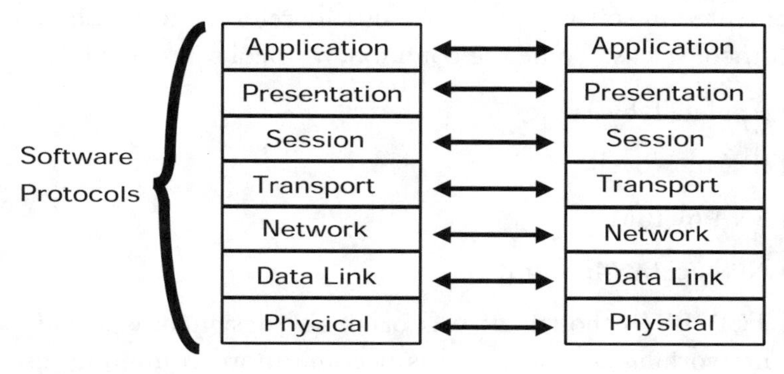

Figure 6.1 The OSI Model.

Table 6.1

Local talk	230 Kbps
Ethernet	10 Mbps
Token ring	4 or 16 Mbps
FDDI	100 Mbps
100 Mb Ethernet	100 Mbps

the types of wires used, the voltages on the wire, and the speed of the information on the wire. Table 6.1 shows the most common interconnect technologies used for LANs and how fast information is moved across the media.

Local talk is supported only by Apple computers and is appropriate only for very small networks. Ethernet is the most common interconnect technology and a greatest number of higher level protocols will run on top of ethernet. Token ring is an aging interconnect technology which was originated by IBM. FDDI and 100 Mb Ethernet are the current state of the art technologies. If you need to run multimedia applications over your network, you should ask your IT manager what the plan is for implementing FDDI or 100 Mb Ethernet.

The top five layers of the communication model are generally implemented in software. They are usually referred to in the industry as "protocol stacks." Some the common protocol stacks for LANs are:

- Appletalk by Apple
- IPX by Novell
- SNA by IBM
- Netbuei by Microsoft
- TCP/IP (Although IP was originally designed as a wide area networking protocol, it has become more commonly used in local area networks).

6.1.2 WANs

A WAN is a Wide Area Network. The nodes of a wide area network can be many miles apart. The public telephone network is an example of a wide area network. Many companies have implemented private WANs with leased lines, microwave links, or satellite links. Traditionally, wide area networks have had much lower bandwidth than LANs. It is much easier to run higher data rates over shorter distances. Table 6.2 will give you an idea of the data rates supported by the most common WAN interconnect technologies.

The protocol stacks most commonly run over a wide area network are TCP/IP and X.25. TCP/IP is the protocol upon which the Internet is based.

6.1.3 Remote LAN Access

Remote LAN access makes it possible for users who are not physically connected to your corporate LAN to gain access to all your LAN resources by way of a WAN connection (Figure 6.2). This merging of LAN and WAN technology is made possible by some of the new lower cost, high-speed WAN interconnect technologies such as ISDN and Frame Relay.

The two primary modalities for remote access are called "remote node" and "remote control." A remote node implementation of remote access makes the remote computer appear as if it is on the

Table 6.2

Pots (plain old telephone service)	28.8 Kbps
Frame Relay	56 Kbps-1.54 Mbps
ISDN (BRI)	128 Kbps
ADSL	6 Mbps downstream, 384 Kbps upstream
T1	1.54 MbpsΩ2.048 Mbps
ATM	1.54 Mbps-155 Mbps

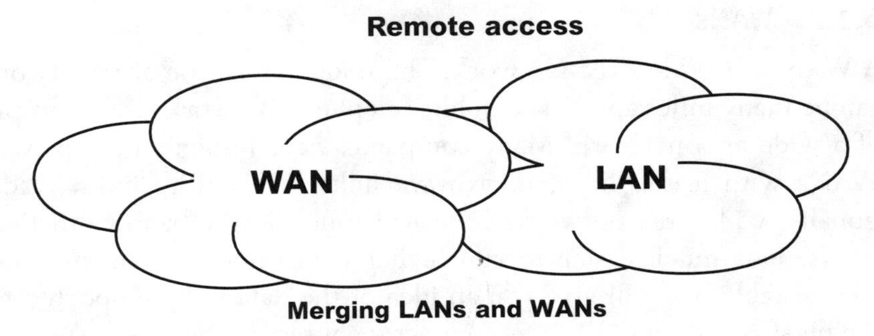

Figure 6.2 Remote access

LAN. A remote control implementation of remote access allows you to log on to a computer that is on the LAN and take control of that computer. Remote control implementations are less transparent to the user but are usually cheaper and easier to implement. "PC Anywhere" is a well known example of remote control software.

Remote access can be implemented by adding a special-purpose device known as a "remote access server" to your network. It provides the hardware necessary to interface WAN connections to your network. There are also vendors such a Microsoft which sell software-only solutions require a designated PC as the remote-access server. You can buy PC plug-in cards to handle the interface to the WAN hardware. Table 6.3 below gives the names and phone numbers of a few remote access server vendors.

Table 6.3

3Com	1-800 638-3266
ACC	1-800 242-0739
Cisco	1-800 553-6387
Gandalf	1-800 426-3253
Livingston	1-800 458-9966

6.2 Which telecommunications technologies facilitate remote-access implementation?

The widening bandwidth of WAN interconnect technology is driving the application of remote access. The decline in the cost of wide area network bandwidth in the late 90s has been dramatic. Figure 6.3 illustrates how the price to performance relationship changed in California in the last few years. Almost 10 times the bandwidth is available for only twice the cost.

Integrated Systems Digital Network (ISDN) and Frame Relay are available in most parts of the United States today. Aymmetric Digital Subscriber Line (ADSL) is in wide spread field trials and should be widely available in five to seven years. The following sections give more detailed description of each of these technologies.

6.2.1 ISDN

Integrated Systems Digital Network lines are all-digital telephone lines designed to carry all types of voice and data traffic over a single pair of wires (hence, the term integrated). In the United States, regional Bell operating companies such as Pacific Bell, Nynex, or Bell Atlantic provide both the service and the installation. It is possible to have both voice and data connections run simultaneously over a single ISDN line. ISDN lines come in two flavors: basic rate and primary rate. Basic rate lines have 128 Kbps of bandwidth divided into two channels. Primary rate lines have 1.54 Mbps of bandwidth divided into 23 channels. Regular analog telephone lines can pass only very

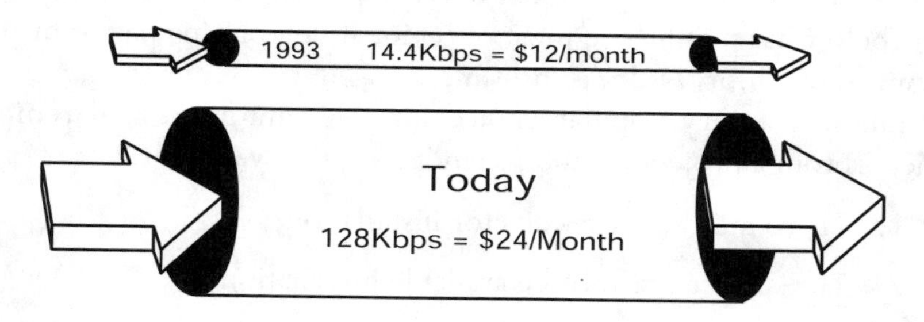

Figure 6.3 The change in bandwidth.

limited signaling (control information). Analog phones can generate only 14 different control signals: offhook, onhook, and the 12 dtmf tones on the keypad. Because ISDN lines are digital, they carry signaling on a separate embedded signaling channel. ISDN devices contain microprocessors and can generate a wide vocabulary of signaling and control information. Because of this expanded vocabulary, ISDN devices can offer a wealth of advanced services. The same circuit can be used to make a voice call, a data call, or a video call.

An important thing to remember about ISDN is that it is a "switched" service. In the 70s and 80s, most wide area data network links were point-to-point leased lines. If you had a videoconference unit at one end of a leased line, the only thing you could connect to was whatever was at the other end of the leased line. With an ISDN line, you can use the public switched telephone network to call anyone else in world with an ISDN line. In most cases, switched services are considered superior because of their flexibility; however, there may be cases where it is desirable to limit a remote worker's connection choices. With an ISDN line on his desk, a remote worker can just as easily dial a networked game service as the corporate LAN. Frame Relay is a better choice for applications where switched service is either not required or undesirable.

In most areas of the United States, ISDN is inexpensive. The cost of a terminal adaptor (the ISDN equivalent of a modem) ranges from $150 to $900. In California, there is a monthly charge of $25 per basic rate line. The per-minute and toll charges are the same as regular analog phone service. Each Bell operating company has a different pricing structure. It is best to check the local Bell operating company's web site to get the most current pricing. The following Web page below has pointers to most of regional Bell pricing pages: http://www.cptech.org/isdn/isdn.html.

ISDN is a very popular choice for telecommuters and remote offices. From a project planning standpoint, be aware that:

1. There may be long waits for installation;
2. The service may not be available internationally;
3. Phone companies are having difficulty training their personnel so the lines are frequently installed incorrectly;

4. User equipment and software is typically difficult to configure.
5. ISDN lines are inoperable in the event of a power loss. (Many people are tempted to disconnect their analog phones once an ISDN line is installed. This is not a good idea because you don't want to lose all your phone service in the event of an emergency.)

Despite these negatives, the increased bandwidth is worth it. The difference between connecting via an ISDN and analog lines is like the difference between driving a car and riding a bicycle. Your people can waste a tremendous amount of time uploading and downloading files over a slow analog link. As a project manager, make sure you have a back-up plan until you see your new infrastructure work. Once the lines are up, they are pretty reliable.

6.2.2 Frame Relay

Frame Relay is another new high-speed WAN interconnect technology. Frame Relay service is a very cost effective replacement for leased data lines. It differs from ISDN in the following ways:

1. It is not a switched service. Frame Relay provides permanent virtual circuits. It is good choice for connecting two remote offices or for giving a telecommuter a high-speed data line which connects only to the corporate LAN.
2. Frame Relay is really a data-only service. It was not designed to support voice services. Some private networks have attempted to implement voice services over Frame Relay, but the complexity is high and the results are usually poor.
3. There are usually no per-minute charges for Frame Relay service. In California, residential Frame Relay service is a flat $150 per month.

In comparing the cost of Frame Relay to ISDN, it is important to understand whether the data use will be continuous or occur in bursts. For applications where a connection is up, and in use eight hours a day, five days per week, Frame Relay will be substantially cheaper because there is no per-minute charge. The table below compares the cost of ISDN and Frame Relay for two different types of users. User 1 is a graphic artist who works at a remote site and needs to upload 20 megabyte files three times a day. User 2 is a customer support engineer who stays connected to the network eight hours day, five days per week. The calculations assume a 6 cent per minute usage charge for ISDN.

User 1. Graphic artist

- ISDN charges = monthly charge + per-minute charge
  = $24 + 4week/month*1.5hour/week*
  60 min/hr*.06cents/minute
  = $45.60/month
- Frame Relay charges = $150/month flat rate

USER 2. Customer Support Engineer

- ISDN charges = monthly charge + per minute charges
  = $24 + 4week/month* 40hours/week*
  60 min/hr*.06cents/minute
  = $600/month
- Frame Relay charges = $150/month flat rate

As you can see, the graphic artist is much better off with the ISDN line and the customer support engineer does better with Frame Relay service.

6.2.3 ADSL

Aymmetric Digital Subscriber Line is the technology that I believe will dominate remote access in the future. Apparently I'm not alone. Microsoft, Intel, and Compac recently announced an alliance to speed the implementation of this particular technology. One potential approach allows users to receive up to 6 Mbps in a downstream channel and send up to 384 Kbps in an upstream channel. ADSL was first proposed by Bellcore in 1989. In January of 1998 the ADSL Forum's Web site listed 39 active field trials. It is hard to guess exactly when ADSL will become widely available. There are a variety of technical, legal, and pricing issues that have yet to be resolved. Many industry leaders believe that ADSL will be widely available in five years. If you are looking for remote access in 1998, your choices are basically analog modems, ISDN, or Frame Relay.

6.3 What types of technology are available to secure my network?

Network security is always a strategic concern. Opening your network to remote workers also opens your network to the possibility of unauthorized access. There are a number of measures a company can use, either alone or in combination, to protect its confidential data.

6.3.1 Dial back

Many remote access servers support a feature called "dial back." This security method requires that the central site be programmed with each remote user's phone number. Users must dial into the server and request that the system call them back. It is not possible for users to dial directly into the network. This type of security can be a problem for nomadic workers, however.

6.3.2 PAP and CHAP

Password Authentication Protocol (PAP) provides basic password validation using a simple two-way handshake at the beginning of a session. PAP is not the most secure authentication method, but it is supported on a wide variety of remote access equipment platforms and provides some security when for operating in a multi-vendor environment.

Challenge-Handshake Authentication Protocol (CHAP) provides additional security by encrypting passwords and utilizing a three-step handshake.

6.3.3 Calling line ID

Many Bell Operating Companies are offering a service which provides the phone number of callers to the receivers of incoming calls. Remote access servers can use this information to accept or refuse incoming calls.

6.3.4 RADIUS

Remote Authentication Dial In User Service (RADIUS) is a security system for client-server environments that uses a single authentication server to provide security services for an entire network. The centralized database offers ease of administration.

6.3.5 Security cards and dynamic password authentication servers

These types of systems are appropriate when a very high level of security is required. Users carry electronic security cards which create unique passwords for each login. The following vendors offer security cards and dynamic password authentication servers:

CryptoCARD, Inc., Buffalo Grove, IL;

Digital Pathways, Inc., Mountain View, CA;

Enigma Logic, Inc., Concord, CA;

Lee Mah DataCom Security Corporation, Hayward, CA;

Security Dynamics, Inc., Cambridge, MA;

Security Corporation, Naples, FL.

6.4 Conclusion

Selecting networking technology that underlies your network is crucial in determining the performance of your communications infrastructure. It is important to balance providing the maximum bandwidth with providing equal access to all team members. Not all communications technologies are available in all areas, particularly if you are working internationally. Remote team members should not be technological second-class citizens. Work closely with your IT manager and document the requirements of all team members.

References

[1] Martin James, *Local Area Network Architectures and Implementations*, Prentice Hall, 1989, p. 95.

7

Implementing a Telecommuting Program

A THRIVING ECONOMY combined with difficulty recruiting and retaining qualified personnel has motivated many companies to adopt flexible work arrangements. Some thoughtful planning is required to assure the change in work arrangements is beneficial to both employers and employees. This chapter will help you understand the different types of telecommuting and give you guidelines for implementing a program in your organization.

7.1 What is telecommuting?

Most managers associate the term telecommuting with full-time home office arrangements. In fact, the term telecommuting is used to describe a variety of flexible work arrangements that allow employees to work at a distance from managers and coworkers. The Gartner Group estimates that by 1999, 80% of all employers will implement some form of telecommuting for about half their employees [1]. Some of the more common types of telecommuting are:

- *Part-time home office.* This is the most prevalent form of telecommuting. Employees perform business functions at home for a designated number of hours per week. However, their primary place of work remains a corporate office. Studies show the average number of telecommuting days per week is 1.5 [2].
- *Full-time home office.* The employer and employee create a primary place of work in the employee's residence.
- *Satellite office.* This is a remote office location usually placed within a large concentration of employee residences, allowing employees to share common office facilities and reduce the time and expense of commuting to the main office.
- *Mobile or nomadic office.* Nomadic telecommuters work from constantly changing work locations. Common examples of nomadic telecommuters are sales people and customer service representatives.

7.2 How do I set up a pilot program?

Like any other project, the success of a telecommuting program depends on a systematic approach to implementation. Before instituting a telecommuting program across your enterprise, we highly recommended you implement a pilot program. Successful programs require the support, cooperation, and involvement from many parts of your organization. As illustrated in Figure 7.1, executive staff, human resources managers, legal staff, information systems administrators, facilities planners, labor unions, managers, coworkers, and the telecommuters themselves are all stakeholders in your telecommuting program. The champion of a telecommuting program will encounter different forms of support and resistance from each of these groups. A pilot program will allow you to build the required

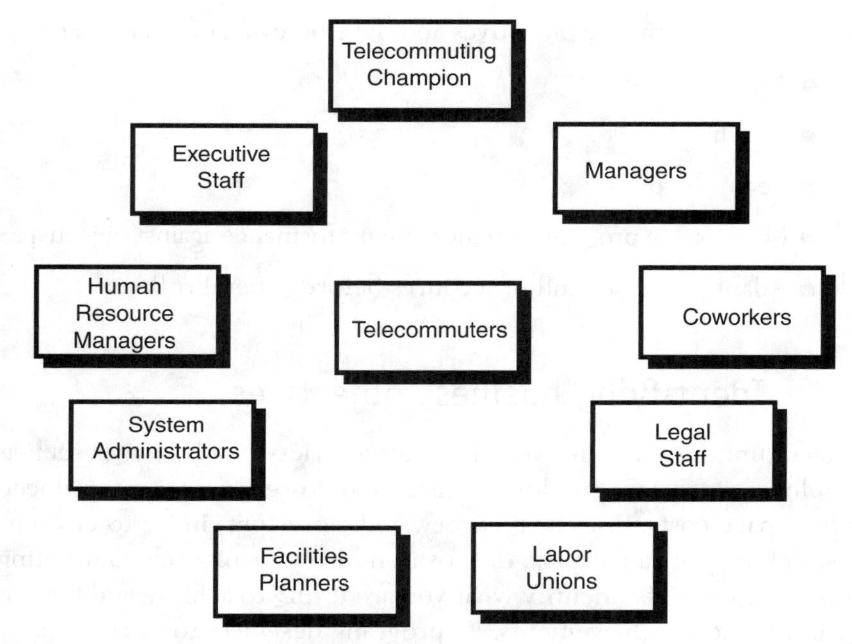

Figure 7.1 Telecommuting stakeholders.

support and test new processes and technology before implementing telecommuting across your enterprise.

Successful telecommuting pilots coordinate the interests of stakeholders with new business processes and technology. The checklist below gives a high-level view of the steps required to set up a pilot program.

- Analyze business processes and identify business objectives.
- Select a champion.
- Interview stakeholders and identify issues.
- Formulate policies and procedures.
- Decide on the technical requirements.
- Analyze cost versus benefit.
- Select the participants.

- Set performance objectives and develop evaluation criteria.
- Train the team.
- Sell the process.
- Begin the pilot.
- Monitor the program and measure performance against objectives.
- Adapt processes and procedures before general rollout.

7.3 Identifying business objectives

Telecommuting can offer your business a variety of advantages such as employee retention, employee satisfaction, lower labor costs, reduced office space costs, disaster recovery, and closer proximity to customers. Before you can analyze the costs and benefits of a telecommuting program, you must identify what you are trying to achieve and how to quantify those achievements. A program designed to retain highly skilled technical employees and reduce turnover will be different from a plan designed to put employees closer to customers. It is important to realize that telecommuting alters work patterns and business processes. Many companies are attracted to the cost savings associated with reducing office space. Studies show that significant workplace changes that are entirely cost driven are much less likely succeed than those that are business driven [3]. I can tell you from personal experience, there is a world of difference between a mandatory and a voluntary telecommuting program. On more than one occasion, MSI has been asked to redesign poorly conceived mandatory telecommuting, space-reduction programs. Many people simply do not want to work at home. They may not have the space or the family support. If your company's program creates the perception that employers are shifting the cost of doing business to the employees, you can expect the most hostile and emotional response to a business change you have ever experienced. Any money saved by reducing office space can be lost to decreased productivity and employee turnover. Successful telecommuting and alternative office programs create modified workplaces that not only reduce cost, but improve overall business processes. An excellent reference on this subject is the book, *Workplace by Design* by Becker and Steele. According to the authors, "In a business-driven workplace, the goal is to become more competitive, not just reduce costs" [4]. The emphasis is shifted from how much money the company saves to how much more can be produce by the same expenditures. Before initiating your pilot telecommuting program, identify how a geographically distributed workforce can improve your business processes and how you can quantify that improvement.

By clearly defining your program's objectives you will be able to demonstrate to both employees and executive staff that the new work style is better, not just cheaper. Becker and Steele suggest developing an "Integrated Workplace Strategy" that exploits the advantages of various workplace settings. The majority of telecommuters work in a corporate facility one or more days a week. Successful businesses have found that telecommuters should not perform the same work that they do from home when they are in the office. It makes more sense for employees to focus on communication and group activities when they are working at the main site. Even if you are not reducing office space as a part of your telecommuting program, it may make sense for you to rethink the layout of the main office. Shifting from rows of isolated cubicles to conference rooms and group work areas may create a more effective work place.

7.4 What types of legal and policy issues should I consider?

Create a written telecommuting policy that all the stakeholders review, understand, and approve. We also recommend a written telecommuting agreement be developed that spells out the specific conditions of each telecommuter's work arrangement. After formulating your policy, developing a form individuals and managers use as a template for a telecommuting agreement is not difficult. The first policy issue to clarify is whether telecommuting is voluntary or mandatory. Most companies implement telecommuting as an employee benefit and program participants view participation as a privilege. Voluntary programs are usually much easier to administrate. There are fewer complaints about reimbursement, fewer performance monitoring problems, fewer insurance claims, and fewer legal actions. However, companies are within their rights to make telecommuting a condition of employment. If you choose to implement a mandatory program, you need to pay particular attention to policy development. Your policy manual is like the rule book for a game no one at your company has ever played before. Without it, the players are guaranteed to disagree on the rules. If you have people who didn't want to play in the first place, you are more likely to wind up in court. The following sections give an overview of the types of issues which should be covered by your policy.

7.4.1 Criteria for selecting telecommuters

Many managers believe only very senior people or team members who have been with company for many years can telecommute successfully. I'm surprised that some companies don't include, "must be able to leap tall buildings in a single bound," as a part of their criteria. I can't tell you how many times I've heard, "You can't just hire telecommuters off the street." The problem is not with potential telecommuters, the problem is with the organization. There is an inverse relationship between requirements for a successful telecommuter and the maturity of the organization (Figure 7.2).

If the processes, tools, and training are not in place, only the supermen and superwomen survive. If the organization is operating at a high level of maturity, junior people can be selected, trained, integrated, and supported. The first step in determining the criteria for selecting telecommuters is assessing your organizational readiness to accept virtual workers. The list below gives specifics to help determine the criteria for selecting telecommmuters.

Aligning telecommuters

Goals

- Has an ROI been done for both the telecommuters and the company?
- Do the telecommuters understand the company's objectives?

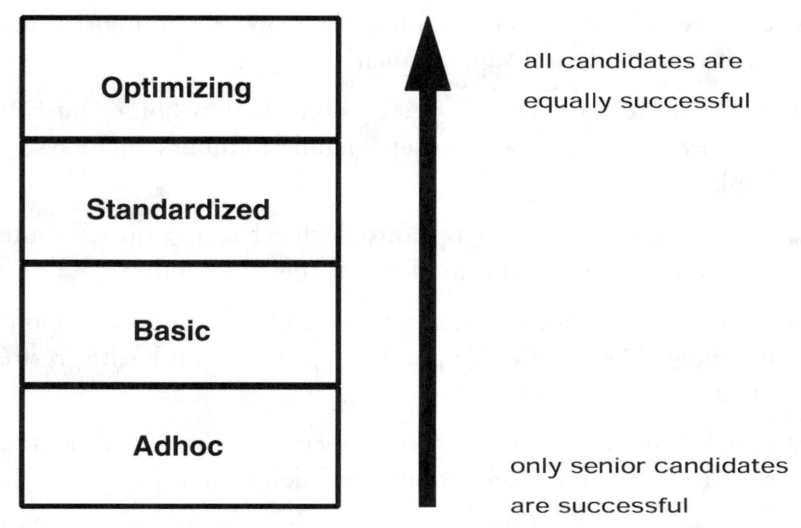

Figure 7.2 The relationship between organizational maturity and the seniority of telecommuters.

- Has child care been arranged?
- Do telecommuters believe working from home is provides a benefit to them?
- Have the telecommuters' families been counseled? Have their expectations been appropriately set?

Process

- Are performance metrics in place?
- Do the telecommuters need mentoring or training? Is there a training plan in place?
- Are the telecommuters completely trustworthy? Is there a plan in place for security including tracking remote equipment, protecting confidential information, securing dial-in computer capabilities, and verifying financial transactions?
- Is there a time-keeping methodology in place for non-exempt workers?

- Can the telecommuters maintain their own equipment? If not, what is the technical support plan?
- Are the telecommuters responsible for at-home business expenses? If not, is there a user-friendly reimbursement system in place?
- Are the telecommuters responsible for backing up computer data? If not, what plan is in place for this function?
- How will the telecommuters access human resources support for things like vacation requests or questions on health, retirement, or other benefits?
- Are the telecommuters's jobs information based or do they require frequent transfers of physical deliverables?

Tools

- Do the telecommuters' jobs require equipment that is unsafe or inappropriate for a home setting?
- Are the required computer programs and data available via remote access?

Skills

- Are the telecommuters proactive in reporting equipment problems?
- How refined are their written and verbal communication skills?
- How experienced are they with the forms of electronic communication your team depends on?
- Are the telecommuters self-motivated, organized, flexible, and able to manage time?
- Do the telecommuters understand distance communication?

Other:

- Are the telecommuters' home office spaces acceptable in terms of things like safety, quiet, and light?

We are often asked, "How can we tell someone we don't think that they are a good candidate for telecommuting without insulting them or making them think the management is playing favorites?" We have found the idea of organizational maturity is very helpful for dealing with this problem. It is not that the telecommuter is somehow inadequate, it is that the organization is not ready to support them yet. There is a big difference between saying, "You can't telecommute, because I know you can't handle keeping your equipment up," and saying, "We don't have a technical support process for telecommuters in place yet, therefore we are starting our telecommuting program with the systems administration group. They are already trained to maintain their own equipment. We are in the process of planning how to support other telecommuters."

7.4.2 Policies for time keeping and availability

Every telecommuting policy should include a section that explains the concept of availability and describes how the standards will be set by the manager and team member. It is not recommended, however, that the actual hours be fixed in the policy as they may vary from person to person. The individual telecommuting agreement is a better place for documenting the specific hours.

If non-exempt workers will be telecommuting, you need a method for submitting time sheets. A non-exempt employee must be paid by the hour, not by the objective. Location has no bearing on the Fair Labor Standards Act. This does not mean that non-exempt employees are poor candidates for telecommuting. As with all employees, a high degree of integrity should be a criteria for selection as a telecommuter. Managers of both exempt and non-exempt workers need to establish and monitor clear performance goals.

7.4.3 Guidelines for reimbursement

Every company must decide its policy for reimbursing home-office expenses. All types of policies have proven to be workable. Mandatory programs generally require a more generous reimbursement program than voluntary programs. It is important to eliminate surprises for both the company and the telecommuter. Inappropriately set financial expectations can poison the relationship between the telecommuter and the employer. The case study below illustrates how crucial it is to have a correct reimbursement policy.

Case Study Bill was a senior programmer for a small high-technology company. His employer had been experiencing rapid growth for the past two years. Office space was in short supply. Since Bill had a 50-minute commute, both he and his manager, Tom, agreed it would be mutually beneficial for Bill to begin working primarily from home. Bill had a strong work ethic and excellent long standing relationships with both his manager and his coworkers. He also had 15 years experience in the networking industry and excellent technical skills. Both Bill and Tom were confident that Bill could install and maintain the ISDN and remote-access equipment required for Bill to do his job remotely. Bill and Tom agreed Bill would select and purchase the equipment he needed and the company would reimburse him for the cost. Tom also agreed that the company would reimburse Bill for telephone bills, office furniture, and office supplies. Bill selected and installed his equipment within three weeks. Bill's productivity increased dramatically working from home. Bill made weekly visits to the office and maintained excellent relationships with his manager and coworkers. At the end of the month, Bill submitted his first expense report. Tom was surprised by how much the phone bill was for one month. He was shocked by the cost of the office equipment and by what Bill thought were legitimate business expenses. Tom approved the expenses anyway and renegotiated with Bill what would be considered reasonable for the following months. This left a bad taste in Bill's mouth, but he eventually agreed. Two weeks later Bill began asking Tom when the expense report would be paid. When Tom thought about it, he remembered that the accounting department was chronically slow with expense checks. (Tom had experienced the delays himself when he submitted travel expenses.) When Tom called accounting he was told nothing could be done to expedite the process. As a matter of fact, since these were not standard accounts and charges he should expect the process to take 60 to 90 days. Bill was furious when he learned he would not be reimbursed for another two months. Despite Tom's profuse apologies, Bill accepted a long-standing job offer from a competitor.

A reimbursement policy should address not only which expenses will be paid for by the company, but also when and how the reimbursement will be made. Situations requiring employees to carry business expenses for extended periods of time are breeding grounds for resentment. In many cases it is wise for the company to put a ceiling on the amount it will pay for certain equipment, services, or installations. For example, a company may be willing to pay for the normal installation of an ISDN line (special high-speed data telephone connection). However, in some instances the new connection may require that a trench be dug in the employee's front yard and the yard resodded. It is best to have thought through these scenarios before beginning your pilot program. Without these types of policies it is impossible for either the company or the telecommuter to do a cost-benefit analysis of the situation. Below are some types of expenses you may want to consider as you develop your reimbursement policy.

Potentially reimbursable expenses

- Computers and office equipment. (Most companies have found it is much cheaper in the long run to purchase standard computer hardware and software than it is to maintain a wide variety of employee purchased equipment.)
- Office furniture
- Travel back and forth to the main office
- Costs for increased expenditures in light, heat, and water bills
- Costs for business-related phone service
- Costs for fire and theft insurance on office equipment
- Costs for upgrading the home to meet safety and electrical standards
- Costs for e-mail and online services
- Costs for installation of high-speed access lines (ISDN, Frame Relay)
- Costs for copying, delivery services, and postage
- Costs for maintenance on equipment

7.4.4 Insurance concerns

If you are putting thousands of dollars of expensive equipment in a telecommuter's home, you should first determine whether your insurance carrier covers off-premises assets. Most homeowners insurance policies need to be modified to cover business equipment. Many renters do not carry any type of insurance. Your policy should clarify who is assuming responsibility for the safety of the equipment. The subjects of workers compensation insurance and liability are covered in section 8.4.10.

7.4.5 Personal access to business equipment

It is awkward to put computers, copiers, and fax machines into homes and completely restrict telecommuters and their families from using the equipment for personal reasons. In some cases, however, for reasons of safety or confidentiality this really is a requirement. In cases where this type of restriction is necessary, it is best to have issue clearly spelled out in your company's telecommuting agreement. Having the written agreement will help the telecommuter explain the situation to the family and deal with their inevitable disappointment.

7.4.6 Safety and environmental standards and home-office inspections

Telecommuters may not be aware of the environmental and safety requirements for home-office equipment. Many homes have old wiring and are not appropriately grounded. In some cases, dusty or humid conditions will damage equipment. Your telecommuting policy should include safety and environmental standards to protect both the employee and the equipment. When telecommuters are highly motivated to work at home, they may be inclined to work in unsafe conditions rather than give up the privilege. A policy that requires home office safety inspections is a prudent idea because the courts hold companies responsible for the safety of employees. A jury in New York awarded a telecommuter and her neighbors a settlement in excess of six million dollars after her inappropriately self-installed business equipment started a fire in her apartment building [5].

7.4.7 Equipment tracking

Your telecommuting policy and telecommuting agreement should specify when and how employer-owned equipment will be returned after the telecommuter's employment is terminated. In most states, it is illegal for an employer to deduct the value of returned equipment from the employees last check. Your company should implement an asset management and tracking system before beginning a telecommuting program.

7.4.8 Security

It is important for telecommuters to understand their responsibilities for safeguarding company information. It is difficult to estimate how much proprietary information is relocated on stolen or lost laptop computers. Developing policies to protect your company's information is not merely for the paranoid. It is a lot easier break in to someone's home to steal competitive information than it is to break in to a well secured corporate office. Chapter 7 discusses various technologies which can be used to improve security; however, the technology isn't adequate without procedures and policies. The most elaborate network login procedure is defeated by users who carry diskettes around in their pockets or misplace printed copies of documents.

7.4.9 Employment law issues

The subject of employment law is complex. Companies should consult their own legal counsel before implementing a telecommuting program. Appendix B contains a list of questions we have been asked in our consulting practice and the responses we received from our employment counsel, Thomas Makris of Rosenblum, Parish and Isaacs, PC. His responses are intended to provide general information and are not legal advise applicable to your particular situation.

7.5 Determining your return on investment

Return on investment needs to be examined from both the employer's and the employee's perspective. Without the proper planning, the telecommunications costs for knowledge workers can be staggering. A single programmer who maintains a dual ISDN connection six hours a day will rack up a monthly bill of $864. When they begin telecommuting, many employees think only of the reduction in transportation cost. That other types of costs may increase (such as utilities costs or costs for out-of-the-home child care) often isn't clear until they receive the first bill. By understanding the business objectives of your telecommuting program (such as reduced turnover or disaster recovery) and doing a detailed analysis, you can appropriately set the expectations of both the employer and the employee. We have included in Appendix D a series of questionnaires that we use to help companies collect the information required to decide between the options of expanding a main office, using voluntary home offices, or opening a satellite office.

When we look at the costs associated with a telecommuting program, we generally separate the recurring costs (such as monthly phone bills) from the non-recurring costs (such as equipment purchases or initial training). Depending on the amount of initial investment, it may require more than one year to recover all of the startup expenses. If your program's objectives have to do with employee retention and productivity increases, costs can generally be projected more accurately than benefits. Fortunately, there is a growing body of literature that documents the type of performance you can expect [2,6]. When we do a return-on-investment analysis for a client we typically present three options.

1. The ROI assuming a conservative benefit. For example, a productivity increase of 10% and reduction in turnover of 5%.
2. The ROI assuming an average benefit. For example, a productivity increase of 25% and a reduction in turnover of 12%.
3. The ROI assuming an above-average benefit. For example, a productivity increase of 35% and a reduction in turnover of 20%.

To get the most accurate projection, we generally use figures that are specific to an industry. Since there are so many studies being published in this area, the information is not difficult to obtain. It probably won't surprise you to find out that if you really scrimp on the startup costs, your productivity gain will be lower. Even assuming the most conservative benefit, your return on investment can be excellent. The case study below documents the return on investment experienced [7].

Case Study Motorola at their Phoenix design center.

One-hundred fifteen telecommuters participated in Motorola's telecommuting program. Users included staff members from the following departments:

- Engineering
- Documentation
- Marketing
- Business services
- Sales
- System and network support

Most telecommuters were in the greater Phoenix area and participated in the program on a part-time basis. However, a number of users were out-of-state, full-time employees. The company's objectives were to provide flexible work options for valued employees, increase productivity, and respond to a local air quality crisis.

Because the users were high technology, "graphics-enabled" knowledge workers it was decided that modem technology was insufficient for the application. ISDN lines were installed at each telecommuter's home. Each telecommuter was provided with an X terminal (typically an older SUN workstation) and an ISDN remote access device. The company paid for the installation of the ISDN line and the continuing access charges. Telecommuters paid for their own home office furniture and similar expenses. Investment was also required at the central site. Cost included installation of ISDN lines, purchase of the remote-access server, and supporting equipment. The startup cost per user was $3,967. The annual recurring cost per user was $1,385.

Productivity for this pilot was measured in the following ways:

1. Existing metrics were used where available (Such as lines of code written per week, document pages edited per day, and cycle time.)
2. Managers were interviewed to report employee performance and MBO results.
3. Employees were interviewed to obtain candid opinions concerning telecommuting effectiveness.
4. Active computer use time was monitored.

All available data indicated an average productivity increase of 12%. Assuming an average employee cost of $80,000 per year, the yearly benefit from the productivity increase alone was $9,600. The startup costs of $3,967 per user were recovered in the first year. For the second and succeeding years, the ROI was 590%.

ROI = (annual benefit/annual nonrecurring cost)*100 -100%:
9600/1385 * 100% -100% = 590%

7.6 What do the employees need to know?

Whether your telecommuting program is voluntary or mandatory, your results will be better if you invest some effort in making certain that participants really understand all the ramifications of working at home or in a satellite office. Most problems arise because the day-to-day reality of telecommuting is different from the fantasy. Work-at-home situations typically require the most education, not only because employees often have unrealistic expectations about mixing work and home, but also because employment law on home work situations is evolving. Telecommuters need to understand how issues of liability, security, privacy, safety, finances, and taxes will be dealt with in a home office situation.

7.6.1 Setting expectations: Dealing with fears, myths, and realities

Even the most enthusiastic potential telecommuters express certain anxieties about being away from the office. The most common fears and myths are listed below, along with techniques for dealing with each issue.

Fear: Telecommuting will result in a loss of recognition by management.
Most telecommuters are concerned that moving to a home office will affect whether or not they are considered for promotions and choice job assignments. Managers need to reassure telecommuters that promotions and job assignments will be given on the basis of merit. A study by Joanne H. Pratt Associates (Pratt, 1993) found that telecommuting does not decrease an employee's chances for promotion. Almost 40 percent of male telecommuters were promoted in 1988 compared to 27 percent of males who did not telecommute. This pattern was also true of women. Thirty four percent of female telecommuters received promotions compared to 21 percent of females who did not telecommute.

Fear: Telecommuters will be left out of office communications.
Managers, telecommuters, and coworkers need to be trained to include all team members in their communications. Chapter 2 gives suggestions for improving communications in a virtual office.

Fear: Telecommuters will be subject to a sweatshop piecework mentality.
Managers should assure telecommuters that objectives will be set based on their past performance in the office.

Myth: Telecommuting is a replacement for child care.
Very few people can care for small children while simultaneously putting in a full day's work. Even if another adult is available in the home to care for children, it may not be possible to maintain a quiet enough environment for the telecommuter to do productive work. It can be very difficult for children to understand that a parent is home but not available to them.

Studies show a much higher rate of success for employees who combine telecommuting with caring for elderly parents or relatives.

Myth: It is always quieter and easier to focus at home.
Veteran telecommuters will tell you that there can be many distractions at home. Many telecommuters find it difficult to get family,

friends, and neighbors to respect their work time and work space. After all, they are just in the next room, so an interruption to ask for help carrying in the groceries or hanging a picture shouldn't be a problem, should it? Some telecommuters can be interrupted often and immediately get back to work. For others it can take 15 to 20 minutes to regain focus. It is very important to set ground rules and get family members to buy into how the new home office situation will work. Telecommuters may find it necessary to ignore their personal phone or the doorbell during working hours.

Even though studies show that on the average, home offices are quieter than corporate offices, new telecommuters are often more disturbed by home noise because they feel responsible for it [5]. A person who would not think twice about raising his voice during a conference call to be heard over construction outside his corporate office may be very distracted and embarrassed by noise made by a child or a pet. Both telecommuters and coworkers need to be trained to regard home office noise as no better or worse than the noise at the main office.

Myth: Telecommuting will reduce my expenses.

Telecommuting will reduce commute expenses; however, other types of expenses may increase. Employees may find themselves heating a house not normally have heated in the winter, going without the benefit of a subsidized corporate cafeteria, and giving up space that they may have previously used for recreational purposes for office equipment. If it is not your company's policy to compensate telecommuters for office space and increased utilities, it is best to have potential telecommuters do an analysis of how their expenses will change once they are working from home.

Myth: My home office won't require much space.

Potential telecommuters often underestimate how much space a home office actually requires. It is difficult to telecommute for more than a few hours per week if you can devote only a corner of a room to work space. Both telecommuters and their families can develop serious misgivings if office equipment and materials slowly take over living space. The next section gives guidelines for setting up a home office.

7.6.2 Setting up a home office

To work productively from home for more than a few hours a week, it is necessary to have a designated office space. Having a space that is out of sight from the rest of the house will help with the psychological aspects of separating work from home. Having a space that is enclosed and lockable is probably necessary for telecommuters with small children or pets. If you keep confidential company material you may want to consider additional security.

New telecommuters often underestimate the amount of table space and storage space required for work materials. Telecommuters need more than just a computer. They need space for manuals, documentation, file cabinets, printers, printer paper, corporate letterhead, fax machines, phones, directories, and so forth. A home office probably requires more space than the office the telecommuter worked in at the corporate location. He or she will need to keep a stock of certain materials and equipment which are kept in a common area in the corporate office

The home office space should conform to guidelines for safety and environmental requirements for equipment. Find a space that is as isolated as possible from noise sources, has sufficient power and light, and maintains reasonable humidity.

Table 7.1
Space to allow for a home office

Desk	15 sq. ft.
Chair	10 sq. ft.
Equipment space for fax, computer, copier and printer	15 sq. ft.
Two file cabinets	15 sq. ft.
Bookshelves	15 sq. ft.
Storage for computer disposables	10 sq. ft.
Room to turn around	20 sq. ft.

If you are telecommuting more than one or two days a week it's hard to work effectively in less than about 100 square feet. The guidelines in Table 7.1 will help you think about how much space you need.

7.6.3 Zoning Laws

Certain local governments may prohibit any type of business office in a home. To avoid any potential liability, companies should assist first-time telecommuters in determining whether any zoning restrictions would apply to their home office. It is a good idea to have renters review their leases for clauses which restrict business use of a home.

7.6.4 The tax consequences of a home office

Tax laws can be complex and vary at the national, state, and local level. In the United States, IRS publication #587, "Business Use of Your Home," explains how home-office tax deductions work. It can be obtained by calling (800) 829-1040. Appendix C contains a list of questions we have been asked in our consulting practice and the answers we received from our CPA, Constance Vandament the president of Vandament Accountancy Corporation in Larkspur, CA. Her responses are intended to provide general information and not tax advise for your particular situation. Employees taking a home-office tax deduction should always consult their tax professional regarding their specific circumstances.

7.7 Conclusion

As technology makes remote access more feasible, the practice of telecommuting will grow in organizations both formally and informally. By formalizing telecommuting policies and procedures, you create a framework which allows all the stakeholders in your organization to understand their roles and contribute positively to the program.

References

[1] Masud, Sam, "Telecommuting: is it for everyone? It might be in 1999", *Government Computer News*, Vol.4, 3 July 1995, p. 41.

[2] Smart Valley Telecommuting Project Final Pilot Results, 25 October, 1995, http://www.svi.org.

[3] Becker, Franklin, and Steele, Fritz, *Workplace by Design*, San Francisco, CA, Jossey-Bass, 1995.

[4] Mahfood, Phillip, *HomeWork*, Probus Publishing, 1992, p 102.

[5] Katz, "Management, Control and Evaluation of a Telecommuting Project: A Case Study," *Information and Management*, Vol. 13, pp. 179-190.

[6] Proceeding of ISDN World Conference, 1996.

8

Conclusion

8.1 How do I put together a plan to move my team forward?

Figure 8.1 shows the steps required to improve your team's effectiveness. The first step in moving your team forward is to assess your team's current maturity level. For small teams, a project manager can do a reasonable job of assessing the team themselves. For larger teams, the assessment may be a project in itself and may require dedicated resources.

Increasing your team's effectiveness requires commitment from all the team members. You should be prepared to devote time to building stakeholder support and setting appropriate expectations. If your team has no experience working virtually, performance may drop before it improves. It is important that your team members understand the process takes time. Just because it isn't perfect two weeks after you've begun isn't a reason to surrender and move everyone back to the main office.

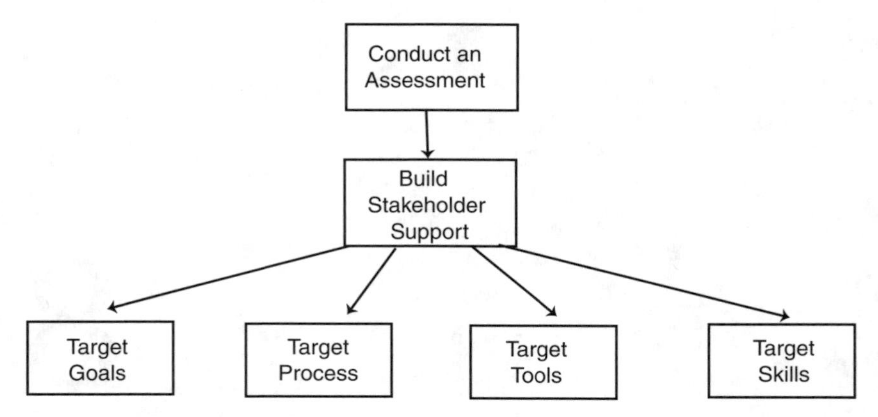

Figure 8.1 Moving your team forward.

Create separate project plans to address the different portions of the alignment model. If you have a large organization, you may want to have subproject managers addressing targeting alignment of the goals, processes, tools, and skills. It is important to keep the four areas moving at a coordinated pace. Use the framework in Chapter 3 to pace your progress.

One subject that we did not cover in this book was the issue of multicultural teams. Fortunately, this subject has been written about extensively and the references in Section 8.3 provide an excellent background on the subject. I can particularly recommend, *Doing Business Internationally: The Guide to Cross-Cultural Success* by Brake and Walker.

Working with virtual teams creates an exciting and rewarding environment. Once you have experienced the new work style it's hard to conceive of returning to the restrictions of traditional work arrangements. Section 8.3 provides a list of additional information resources you may want to utilize in implementing your team.

8.2 Some final thoughts

The writing of this book has been yet another experience in working on a virtual team. My editors and reviewers have excellent collaborators on this project and were of great assistance in clarifying the publishing process to a new author. My respect and gratitude for authors that I have so leisurely read and referenced has grown immensely. You never know how hard it is to do something until you try it. I have to admit, it is a lot easier to write software.

I hope the information in the last seven chapters has been useful to you. My experience working in the virtual world has been incredibly positive. Feel free to drop into my virtual office and tell me about your experiences. You know where to find me: mhaywood@mgtstrat.com.

8.3 Where can I find more resources?

E-mail references

General overview books

The E-mail Frontier, Emerging Markets and Evolving Technologies, Daniel J. Blum and David M. Litwack, Addison-Wesley, 1994.

Electronic Mail, Jacob Palme, Artech House Publishers, 1995.

E-mail: Achieving Local and Global Communications, Computer Technology Research Corp., 1995.

The Internet Message: Closing the Book with Electronic Mail, Marshall T. Rose, Prentice-Hall, 1993.

PGP: Pretty Good Privacy, Simson Garfinkel, O'Reilly & Associates.

Books primarily about social effects of e-mail

The Network Nation, Starr Roxanne Hiltz and Murray Turoff, Addison-Wesley, 1978, MIT Press, 1993.

Connections: New Ways of Working in the Networked Organization, Lee Sproull and Sara Kiesler, MIT Press, 1991.

Style and Ethical Aspects

The Elements of E-mail Style, David Angell and Brent Heslop, Addison-Wesley, 1994.

The Smiley Dictionary, Seth Godin, Peachpit Press, 1993

Smileys, David W. Sanderson, O'Reilly and Associates, 1993.

Teambuilding references

Managing to Have Fun, Matt Weinstein, Fireside Books, New York , NY, 1996.

1001 Ways to Reward Employees, Bob Nelson, Workman Publishing, New York, NY, 1994.

International references

A Manager's Guide to Globalization, Stephen Rhinesmith, Irwin Publishing, 1996.

Comparative Management: Organizational and Cultural Perspectives, Stanley Davis, Prentice-Hall, 1971.

Doing Business Internationally: The Guide to Cross-Cultural Success, Terence Brake, Danielle Medina Walker, and Thomas Walker, Irwin Publishing ,1995.

Global Management Desk Reference: 151 Strategies, Ideas and Checklists from the World's Most Successful International Companies, Shirley Dreifus, McGraw-Hill, 1992.

Global Solutions for Teams: Moving from Collision to Collaboration, Sylvia B. Odenwald, Irwin Publishing, 1993.

Going International: How to Make Friends and Deal Effectively in the International Marketplace, Lennie Copeland and Lewis Griggs, Willard Books, 1993.

Managing Globally: A Complete Guide to Competing Worldwide, Carl Nelson, Irwin Publishing, 1994.

Managing International Teams, Nicola Phillips, Irwin Publishing, 1994.

International Dimensions of Organizational Behavior, Nancy Adler, Kent Publishing, 1986.

Total Global Strategy: Managing for Worldwide Competitive Advantage, George S. Yip, Prentice-Hall, 1992.

Books about particular countries or regions

Doing Business in Mexico, Jay and Maggie Jessup, Prima Publishing, 1993.

The Executive Guide to Asia-Pacific Communications, David L. James, Kodansha International, 1995.

Japanese Etiquette and Ethics in Business, Boye De Menthe, NTC Business Books, 1988.

Korean Etiquette and Ethics in Business, Boye De Menthe, NTC Business Books, 1988.

Management in Socialist Countries: USSR and Central Europe, Witold Kiezun, Walter de Gruyter, 1991.

Networking References

Distributed Systems Management, A. Langsford and J.D. Moffett, Addison-Wesley Publishing Company, 1988.

Frame Relay Networks: Specifications and Implementations, Uyless Black, McGraw-Hill, 1994.

Frame Relay Principles and Applications, Phillip Smith, Addison-Wesley Publishing Company, 1993.

Handbook of Computer Communications Standards, William Stallings, Howard W. Sams and Company, 1989.

ISDN: An Introduction, Howard W. Sams and Company, 1989.

Inside Appletalk, Gursharan S. Sidhu, Addison Wesley Publishing, 1990.

Local Area Networks Architectures and Implementations, James Martin, Prentice Hall, 1995.

The X Series Recommendations: Protocols for Data Communication Networks, Uyless Black, McGraw-Hill, 1991.

Internet Companion, Tracy LaQuey, Addison Wesley, 1993.

Networking Standards: A Guide to OSI, ISDN, LAN and MAN Standards, William Stallings, Addison Wesley, 1989.

TCP/IP: Running a Successful Network, K. Washburn and J.T. Evans, Addison-Wesley Publishing Company, 1988.

Telecommunications Protocols and Design, Joseph Hammond and Krzysztof Pawlikowski, Addison-Wesley Publishing Company, 1991.

Telcommuting references

Creating a Flexible Workplace: How to Select and Manage Alternative Work Options, B. Olmsted and S. Smithe, AMACOM, 1989.

Home is Where The Office is, A. Bibby , Headway, Hodder and Stoughton, 1991.

Making Telecommuting Happen, a Guide for Telemanagers and Telecommuters, Bruce Brown, Van Nostrad Reinhold, 1995.

Myths and Realities of Working at Home: Characteristics of Home-Based Business Owners and Telecommuters, J.H. Pratt, 1993.

Telecommuting: A Managers's Guide to Flexible Work Arrangements, Joel Kugelmass, Lexington Books, 1995.

Telecommuting, Osman Eldib, and Danial Minoli, Artech House, Inc., 1995.

The Telecommuter's Handbook: How to work for a Salary Without Ever Leaving the House, Brad Schepp, Pharos Books, 1990.

Telecommuting in Large Organizations, T. Miller, LINK Resourses, 1992.

Telework: Towards the Elusive Office, U. Huws, Wiley, 1990.

Teleworking Explained, G. Gordon, M. Gray, N. Hodson, Wiley, 1993.

Online resources

EMA - Electronic Messaging Association Web site: http://www.ema.org/ema/.

Management Strategies, Inc Web site: http://www.mgtstrat.com/.

Smart Valley Web site: http://www.svi.org.

Pacific Bell Telecommuting Resource Guide: http://www.pacbell.com/Lib/TCGuide/index.html.

Telecommuting and Telework Info Page: http://grove.ufl.edu/~pflewis/commute.html.

Telecommuting Advisory Council: http://www.telecommute.org.

Telecommute America: http://www.att.com/Telecommute_America.

Virtual office references

Agile Competitors and Virtual Organizations, Steven L Goldman, Rodger N. Nagel and Kenneth Preiss, International Thomson Publishing Inc., 1995.

The Digital Workplace, C.E. Grantham, Van Nostrand Reinhold, 1993.

Going Virtual, Ray Grenier and George Metes, Prentice-Hall, 1995.

Knights of the Tele-Round Table, Jaclyn Kostner, Time Warner, 1994.

Paradigm Shift: The New Promise of Information Technology, Don Tapscott and Art Caston, McGraw-Hill, 1993.

The Virtual Community: Homesteading on the Electronic Frontier, Howard Rheingold, Addison-Wesley, 1993.

The Virtual Corporation Structuring and Revitalized the Corporation for the 21st Century, William Davidow and Michael s. Malone, Harper Business, 1992.

Workplace by Design: Mapping the High Performance Workscape, Franklin Becker and Fritz Steele, Josey-Bass Inc., 1995.

Connections: New Ways of Working in the Networked Organization, Lee Sproull and Sara Kiesler, MIT Press, 1991.

Computing Strategies for Reengineering Your Organization, Cheryl Currid, Prima Publishing, 1996.

Appendix A: Assessment checklists

THIS APPENDIX PROVIDES checklists with specific considerations for assessing various types of distributed team members. The checklists are to be used in addition to the high-level checklists provided in Chapter 3.

A.1 Assessing consultants and third-party organizations

Consultant: Individual contracted for project-specific work.
Third-party developer: An outside company that will develop part or all of a product, usually in a "turn-key" fashion.

Goals

- Is this really their main line of work?
- Are they paid by the hour or on completion of objectives?
- What are the consequences to the consultant or third party organization if objectives are not met?

Process:

- Will they meet your aggressive schedules?
- Will they participate in concrete detailed planning?
- Will they produce status reports as necessary?
- Will they attend team meetings if required?
- Will they produce documentation that flows easily into your team's corporate memory?
- What is their attitude toward management's role?
- Does your payment and accounting process meet their terms? (Expect poor performance if you pay outside parties slowly.)

Tools:

- What systems or tools do they have in place for corporate memory?
- Does your company own, or have permanent immediate access to all the tools used by this outside organization? If not, how certain are you that the work will not have to be modified in the future?
- Are they willing to use your team's project management tools?

Skills:

- What is their experience with your product, industry, or technology?
- How refined are their written and verbal communication skills?
- How experienced are they with the forms of electronic communication your team depends on?
- Do they have experience working on CONTRACT?

Stability:

- Are they able to support you throughout the project at an appropriate level?

- Will they be available to provide long-term support and answer questions after the contract has completed?

A.2 Assessing team members at other corporate sites

Goals:

- What are the business goals of the remote site?
- To what extent does the management of the remote site share responsibility for your objectives?
- To what extent do the team members at the remote site share responsibility for your objectives?
- What is your level of influence on the team members at the remote site? Do they report to you directly? Do they report to another manager at the remote site?
- What cultural differences may affect the motivations of the team members?

Process:

- Is there a mechanism for arbitrating process differences?
- Are performance metrics in place?
- Does the does the team member need mentoring or training? Is there a training plan in place?
- What type of technical support is available at the remote site? Does it support your team's tools?

Tools:

- Is there a mechanism for arbitrating tools differences?
- What is the state of tools integration?

Skills:

- What is their experience with your product, industry, or technology?
- How refined are their written and verbal communication skills?
- How experienced are they with the forms of electronic communication your team depends on?

A.3 Assessing OEM partners

An Outside Equipment Manufacturer (OEM) partner is an organization which manufactures a subsystem which was developed specifically for incorporation in your company's product.

Goals:

- What are the OEM partner's specific numeric sales goals?
- What is their attitude toward partnering?
- Are they looking for an exclusive relationship now or in the future?
- Who are their other customers?
- What are their policies and attitudes about customer support?
- What are their future product plans?
- What are their feature priorities?
- What are their release schedules?
- What is their level of stability?

Process:

- What systems are in place to transfer manufacturing of this entity to another supplier, in the event the OEM partner can not maintain supply?

Skills:

- What experience level do you have with each other's technology?

A.4 Assessing OEM vendors of off-the-shelf hardware and software

Organizations that supply hardware or software system-type components such as operating systems and development tools.

Goals:

- What are their other customers?
- What are their policies and attitudes about customer support?
- What are their future product plans?
- What are their feature priorities?
- What are their release schedules?
- What is their level of stability?

Skills:

- What is their track record for on-time release?
- What is their track record for product quality?

A.5 Assessing vendors

Vendors providers commodities or standard components.

Goals:

- How can they provide you assurance of supply?
- What are their policies and attitudes about customer support?
- What are their future product plans?
- What are their feature priorities?

Skills:

- What is their surge capacity?
- What is their degree of creativity and innovation?

- What is their track record for on-time release?
- What is their track record for product quality?

A.6 Assessing telecommuters

Goals

- Has an ROI been done for both the telecommuters and the company?
- Do the telecommuters understand the company's objectives?
- Has child care been arranged?
- Do telecommuters believe working from home is provides a benefit to them?
- Have the telecommuters' families been counseled? Have their expectations been appropriately set?

Process

- Are performance metrics in place?
- Do the telecommuters need mentoring or training? Is there a training plan in place?
- Are the telecommuters completely trustworthy? Is there a plan in place for security including tracking remote equipment, protecting confidential information, securing dial-in computer capabilities, and verifying financial transactions?
- Is there a time-keeping methodology in place for non-exempt workers?
- Can the telecommuters maintain their own equipment? If not, what is the technical support plan?
- Are the telecommuters responsible for at-home business expenses? If not, is there a user-friendly reimbursement system in place?

- Are the telecommuters responsible for backing up computer data? If not, what plan is in place for this function?
- How will the telecommuters access human resources support for things like vacation requests or questions on health, retirement, or other benefits?
- Are the telecommuters's jobs information based or do they require frequent transfers of physical deliverables?

Tools

- Do the telecommuters' jobs require equipment that is unsafe or inappropriate for a home setting?
- Are the required computer programs and data available via remote access?

Skills

- Are the telecommuters proactive in reporting equipment problems?
- How refined are their written and verbal communication skills?
- How experienced are they with the forms of electronic communication your team depends on?
- Are the telecommuters self-motivated, organized, flexible, and able to manage time?
- Do the telecommuters understand distance communication?

Other:

- Are the telecommuters' home office spaces acceptable in terms of things like safety, quiet, and light?

Appendix B

What follows is a list of questions we have been asked in our consulting practice and the responses we received from our employment counsel, Thomas Makris of Rosenblum, Parish and Isaacs, PC. His responses are intended to provide general information and are not legal advice applicable to your particular situation.

1. In situations where employees are telecommuting and their home office is the primary place of work, does having a specific space in their homes designated as office space help limit the company's liability? For example: if they cut themselves making a sandwich in the kitchen during work hours, can they make a workers' compensation claim or sue the company? What if they slip and fall while taking a business call on their cellular phone by the pool? What if they take their laptop to the beach or a coffee shop to do their work and are injured?

   Designating a home office space generally will not limit exposure to workers' compensation claims. If the employee is actually performing services, for example, using the cell phone at the pool, any injury would be considered work related. The same is probably true if the employee is working on a laptop at the beach. If the employee is attending to incidental matters of personal comfort, convenience or necessity, such as making a sandwich during work hours, the injury is still compensable under workers' compensation.

2. If family members other than the employee are injured while in a home office what, if any, liability would the company have? Would having a telecommuting agreement which specifically states that family members should stay out of the home office space and away from home office equipment provide any protection?

    From a liability perspective, a telecommuting agreement requiring the employee keep family members out of the home office provides little or no protection and could arguably increase liability. If employees let their children play in the home office, they may be violating company policy but are still acting within the course and scope of their employment. The company still has potential liability for negligence under the doctrine of respondent superior. The fact that the company has required the employee to sign an agreement saying that family members will not be allowed into the home office may well increase the likelihood of a jury concluding that the employer was negligent in letting it happen.

3. If a company determines that the home office space an employee, or potential employee, has available does not meet company standards for safety or suitability, is the company its rights to terminate or not hire that employee?

    The answer to this question is generally, yes, a company can refuse to employ a telecommuting employee who does not have adequate facilities for effective telecommuting. However, as with any employment-related decision, the company's policies should be evaluated so they are based on legitimate business objectives, that they do not unnecessarily or unreasonably invade the employees privacy interests, and that they do not have the intended or unintended affect of adversely impacting any protected group such as race, religion, or gender, among others.

4. Does a telecommuting agreement that requires home office inspections violate any employee rights?

   Requiring some inspection of the home work space is certainly appropriate. For example, if the company is providing high-priced computer equipment, the company has a legitimate interest in insuring that the equipment is installed properly in an environment that is not likely to damage the equipment. Again, the inspection should be limited to issues which are of legitimate business concern to the company and caution should be taken to avoid unnecessary invasion of privacy or establishing requirements which may adversely impact protected groups.

5. When a telecommuting employee resigns, what recourse does the company have if the employee does not return company equipment? For example, can the company withhold the employee's last paycheck? What if the value of the equipment exceeds their last check?

   As with all the issues we are discussing, the answer depends on which state law you are applying. In California, the labor code allows an employer to deduct the value of equipment that the employee fails to return, but only if the employee has given prior, written authorization for the deduction. The employer cannot deduct anything from wages for equipment lost or damaged through ordinary negligence.

6. Will setting designated home office hours help limit a company's liability?

   Setting designated work hours may or may not be helpful. If the employee is injured at home while working on company business outside of designated work hours, the employee will have a workers' compensation claim. If the employee is injured at home while taking care of personal matters, having designated work hours may reduce the risk of workers' compensation exposure if the injury occurs outside of the designated work hours, but may increase the likelihood of a workers' compensation claim if the injury occurs during the designated work hours.

7. Is it legal for employers to require employees to allow unannounced visits by supervisors or other employees?

This is absolutely a bad idea. This risk of violation of legitimate privacy interest through unannounced inspections is too great. The business justification for requiring unannounced inspections is minimal. The bottom line is that if you think you may need unannounced inspections to check up on an employee, you do not trust your work force well enough for a telecommuting arrangement to work.

8. Do publishing safely guidelines and standards help limit a company's liability? Is the employer responsible if the employee disregards the guidelines?

The reason for publishing safety guidelines and standards is to reduce the likelihood of injury. The question of liability is definitely a secondary issue. If the employee is injured at home while working, the employee is going to have a valid workers' compensation claim whether or not safety guidelines have been published. If a third party is injured in the home office because of something the employee has done that is in violation of safety guidelines, the company's exposure to a negligence claim may be increased. However, this is true in any setting where safety is a question. Safety guidelines always increase the risk that if they are violated, they can be used against the company in third-party claims. Companies still publish safety guidelines because the best defense is to do everything reasonable to prevent the injury in the first place.

Appendix C: The Tax Consequences of a Home Office

What follows is a list of questions we have been asked in our consulting practice, and the answers we received from our CPA, Constance Vandament, the president of Vandament Accountancy Corporation in Larkspur, California. Her responses are intended to provide general information and not tax advice for your particular situation. Employees taking a home-office tax deduction should always consult their tax professional regarding their specific circumstances.

C.1 General questions

1. As an employee, under what circumstances do I qualify for a home office deduction?

   Internal Revenue Code Section 280A, states that to qualify for a deduction, you must use your home office on a regular *basis, and it must be used* exclusively *as your principal place of business. In addition, as an employee, the exclusive and regular use must be for the* convenience of your employer. *You should have your employer issue a statement in writing that your home office is for the employer's convenience.*

2. Does a hoteling arrangement disqualify me from the home office deduction?

   Assuming the tests of Section 280A are met (regular and exclusive use), and assuming that the preponderance of your work is done at the home office, occasional use of your employer's central site should not disqualify your for the deduction.

3. I cannot dedicate a whole room to my home office. Can I deduct a portion of the room?

   Yes, if that portion of the room meets the "exclusive use" test. If it is sometimes used for personal purposes, such as a guest room, it will not qualify.

4. My wife does not work, however, since my employer forces me to work at home and it is too noisy to work with a two-year old and a four-year old in my two-bedroom house, I have incurred additional expenses for outside daycare. Are these expenses deductible?

   No. There is a child care credit available for payments for child care so that you and your spouse can work. If your wife doesn't work, the credit will not apply.

5. I live alone. Because my employer requires me to work at home, I must now heat my house all day long and my utility bills have tripled. My home office is only 20% of my house. Am I entitled to deduction of only 20% of this increase?

   Although the percentage method is the most common, you may use any "reasonable method" to calculate the costs of the home office. If you can substantiate the fact that the utility increase is directly related to the use of the home office, you should be able to deduct the higher amount.

6. I have to store manuals, paper goods and equipment in my basement and garage. Am I allowed a deduction for this and if so how would I calculate it?

   If the storage area meets the "exclusive use" test, the same method used in calculating the home office area will apply. (A new law allows a deduction for storage of inventory and product samples, even if the use is not exclusive, but this applies only if your trade or business is wholesale or retail selling of products.)

7. Since my main office is in my home can I deduct auto expenses when I travel to the main office?

   Yes. For 1998, the allowed mileage rate is 32.5 cents per mile.

8. Is any portion of my bathroom or kitchen deductible?

   No.

9. How does recapture of depreciation work when I sell my house?

   Depreciation on your home office will decrease the cost basis of your house, and will therefore be subject to "recapture" when you sell. Since your home office is now considered to be commercial property, if you sell your house you will not be able to defer tax on the home office portion by buying another home.

   If you no longer have the home office at the time of the sale, all that will affect you will be the reduction of basis by depreciation taken earlier.

10. I used to park my car on the street during evenings and weekends. Now that I work at home, I must rent a parking space during the day for my car because my neighborhood has two-hour meters on weekdays. Can I deduct any of this expense?

    The Internal Revenue Service would consider this a personal expense, not deductible.

11. The telephone company had to dig a trench in my front yard to install my ISDN line. It will cost me $2,000 to re-sod my lawn. Is this deductible?

Probably not, unless you could allocate a portion of the new lawn only to the specific area where the ISDN line is located. The more conservative approach would be to simply add the $2,000 to the cost of your home for deduction later on when the house is sold.

12. I have a second home in Tahoe. Can I convert that to 100% office space?

    If the home is used exclusively and regularly *for business, yes.*

13. I've decided to rent a separate office space because my employer doesn't provide a work space. Is this deductible?

    Yes, if exclusive and regular use requirements are met and if it is for your employer's convenience.

14. Can I use my van, RV or boat as my home office and deduct it?

    Yes, again if exclusive and regular use requirements are met and if it is for your employer's convenience.

15. How do I deduct my home-office expenses on my income tax return?

    Since you are an employee, you will deduct these expenses on Schedule A, as a miscellaneous itemized deductions.

C.2 Specific deductions

Given that that I qualify for a home office deduction and have found and allocated the space, which of the following expenses are deductible?

A. The cost of office furniture, drapes, carpets, plants, and art for the office.

   Office furniture, drapes, carpets are depreciated under MACRS (modified accelerated cost recovery system). However, in 1997 you may elect to deduct up to $18,000 worth of equipment without depreciating it (Code Section 179). Plants should be fully deductible, but art could be challenged if the cost is excessive.

B. The cost of office equipment such as a fax machine, personal copier, answering machine, backup hard drives, or CD-ROM drives.

These are depreciable or deductible, to the extent of business use. If they are appropriate and helpful to your work, such items are allowable.

C. The cost of business telephone expenses such as cellular phone, voice mail, and pagers.

Deductible. Remember that you may not deduct you home telephone costs except to the extent of specific business calls.Your second business line is fully deductible.

D. The cost of online services.

Deductible, assuming this is essential to your business activity.

E. The cost of business software such as contact mangers, presentation software, and so forth.

Those expenses are deductible.

F. Reference material and subscriptions.

Yes, deductible.

G. Office supplies such as pens pencils paper, staples, fax paper, and so forth.

Those expenses are deductible.

H. Postage

Those expenses are deductible.

I. (1) Utilities bills - electric, gas, water, and garbage
(2) Housecleaning and maintenance
(3) Fire and earthquake insurance
(4) Rent or mortgage interest
(5) Property tax
(6) Depreciation of building

These are all deductible to the extent of the home office use. The normal method of allocating home office expenses is as follows:

Divide the area used for business by the total area of your home. Usually this means using square feet, although it is sometime acceptable to use the number of rooms. For instance, it might be one room out of ten or 100 square feet out of 1,000, giving an allocation of 1/10 or 10% for business

Depreciation of a home office is over 31.5 years. Only the portion of the home's value allocated to the building can be depreciated, not the value of the land.

J. Repairs

Repairs affecting the entire building (such as a roof repair, furnace repair, or exterior paint) must be allocated as above. Repairs to your office equipment would be fully deductible. No deduction is allowed for lawns or landscaping (except as addition to your home basis).

K. Upgrades made only because of home-office situation. For example: upgrading home wiring to include grounding, installation of air conditioning, or sound proofing.

Items which directly affect the office area (such as shelving, sound proofing, paint, or wiring) are deductible or depreciable as a business expense. But upgrades which benefit the entire building are added to your overall cost basis, and only be depreciated the same as the building.

Appendix D: Departmental Return on Investment Questionnaires

This appendix contains a series of questionnaires we have used to collect the data required for a company to analyze and compare the return on investment of a satellite office and a part-time telecommuting program. The questionnaires are to be filled out by different functional organizations within the company and returned to the telecommuting coordinator or champion for analysis.

Facilities questionnaire

Please enter the following costs on an annual, per-user basis. If the information is not available on a per-user basis, entire facility costs are acceptable. If you are entering entire facility costs, please follow the cost with a /F.

Office space costs:

Main office

Building rental; or	________________
Mortgage and depreciation	________________
Property taxes	________________
Fire and earthquake insurance	________________
Building maintenance	________________
Landscaping	________________
Parking	________________
Janitorial service	________________
Building security/parking security	________________
Cafeteria/vending	________________

Are there facilities to support additional personal at your existing main office?________________

Would construction or leasing of additional facilities at the main office be required to support the number of people who would otherwise be in the satellite office?________________
(If yes, an additional questionnaire will be required.)

Will you be able to reduce your current office space if a satellite office or telecommuting program is implemented?________________
(If yes, an additional questionnaire will be required.)

Satellite office

Site selection	________________
Office design	________________
Improvement/remodeling costs	________________

Building rental ______________
Purchase costs ______________
Mortgage Depreciation ______________
Closing costs ______________
Property taxes ______________
Fire and earthquake insurance ______________
Building maintenance ______________
Landscaping ______________
Parking ______________
Janitorial services ______________
Building security/parking security ______________
Cafeteria/vending ______________

Home office

Safety inspection ______________
Electrical upgrades ______________
Office design ______________
Office space reimbursement ______________

Utilities costs

Please enter the following costs on an annual per-user basis. If the information is not available on a per-user basis, entire facility costs are acceptable. If you are entering entire facility costs please follow the cost with a /F.

Main office

Electricity ______________
Water/sewer ______________
Garbage ______________

Satellite office

Electricity ______________
Water/sewer ______________
Garbage ______________

Home office

Enter only if the employee is provided reimbursement.

Electricity ____________________
Water/sewer ____________________
Garbage ____________________

Finance questionnaire

Property taxes

Please enter the property tax rate for both the main facility and the proposed satellite facility:

Main facility ____________________
Satellite facility ____________________

Are there local business taxes at either facility?

Main facility ____________________
Satellite facility ____________________

Is there a payroll tax at either facility? (For example, San Francisco collects a percentage of all payroll within the city limits.)

Main facility local payroll tax rate ____________________
Satellite facility local payroll tax rate ____________________

Human resources questionnaire

Is telecommuting voluntary at your company?__________

Do you have a written telecommuting policy?__________

Do you have a written telecommuting agreement?______________

Will you have formal criteria for selecting telecommuters?________

Is there a plan for training managers, telecommuters, and coworkers in virtual office skills?_______________
If so, what is the cost per employee?_______________

Will there be a policy for reimbursing telecommuters for home office expenses?_______________
If so, indicate the amount per month for each item:

Utilities: heat, electricity, water _______________
Office space _______________
Office supplies _______________

Do you plan to implement minimum home-office safety standards?

____

Do you plan to implement a home-office inspection program to guarantee the safety of the home workplace? _______________

Will it be the responsibility of the company or the employee to pay for upgrades required by safety standards (for example, electrical grounding)? _______________

Will an office-space-reduction program be implemented in conjunction with your telecommuting program? For example, will part-time telecommuters share offices at the main facility?_______________

Will hourly employees be telecommuting?_______________
If so, what will be the cost of implementing a timekeeping program?

____

What is your current average workers' compensation cost?

____

Do you have a quote for workers' compensation in the satellite facility?________________

Do you have a quote for workers compensation with your telecommuting program?________________

How will you provide HR support to a satellite facility? Will an HR be on site part time? Will workers at a satellite office be required to travel for HR support? Will support be provided over the Internet? Can you estimate the cost of your support plan?________________

What is the average rate of absenteeism at your company?__________

What is the average rate of turnover at your company?____________

Does your company currently have performance metrics in place?

____

What is the average burdened cost of an employee at your company?

____

What are your average recruiting costs per employee?

____

What are your average training and development costs per employee, per year?________________

Is the cost of labor significantly less in the location of your satellite office?________________
If so, please estimate as a percentage of annual salary.______________

Does your company provide on site benefits such as a subsidized cafeteria or on-site child care which will not be utilized by remote workers?________________
If so, estimate any cost savings per employee________________

Is telecommuting part of your compliance program with the American's With Disabilities Act?________________
If so, what would be the cost of an alternative form of compliance?

____

Is your company implementing a program to comply with the 1990 Clean Air Act? ________________

If so, what would be the cost of an alternative form of compliance?

____

Information technology questionnaire

Please enter the following costs on a per-user basis for each office arrangement

Voice Communications

Information in this section refers to costs associated with voice communications. If LAN connectivity or data communications is provided by analog modem do not enter that information in this section, enter it the "Data Communications" section. It is HIGHLY RECOMMENDED that the same analog phone line not be used for both voice and data communications.

Analog phone service

Specify the number of analog phone lines per user, the installation cost, and fixed monthly charges
(Local and long-distance usage rates will be entered elsewhere.)

Main office: # of lines: __________
installation cost per line: __________
monthly charges: __________

Satellite office: # of lines: __________
installation cost per line: __________
monthly charges: __________

Home office: # of lines: __________
installation cost per line: __________
monthly charges: __________

Will satellite office users have analog phone service at both the central office and the satellite office?__________

Will home-office users have analog phone service at both the central office and the home office?__________

Specify the *average local usage* charge on a per-minute and per-user basis:

Main office: charge/min: __________
average minutes/user: __________

Satellite office: charge/min: __________
average minutes/user: __________

Home office: charge/min: __________
average minutes/user: __________

Specify the *average long-distance* usage charge on a per-user basis:

Main office: __________
Satellite office: __________
Home office: __________

Specify the cost for the telephone instrument:

Main office: __________
Satellite office: __________
Home office: __________

Voice features

Voice features refer to voice-application features such as call forwarding and call waiting.

Do you use PBX service or Centrex service at your main office?

____

For Centrex: installation cost per line: __________
monthly charges: __________

For PBX service: installation cost per line: __________
monthly charges: __________

Will you use PBX service or Centrex service at your Satellite office?

____

For Centrex: installation cost per line:________
monthly charges:________

For PBX service: installation cost per line:________
monthly charges:________

Will voice-application features be used at the home office?
____

installation cost per line:________
monthly charges: ________

Voice mail

Main office: installation cost per line:________
monthly charges/maintenance:________

Satellite office: installation cost per line:________
monthly charges/maintenance:________

Home office: installation cost per line:________
monthly charges/maintenance:_______

Will satellite-office users have voice mail service at both the central office and the satellite office?________________

Will home-office users have voice mail service at both the central office and the home office? ________________

Pager service

Main office: installation cost per line: ________
monthly charges/maintenance: _______

Satellite office: installation cost per line: ________
monthly charges/maintenance: _______

Home office: installation cost per line: ________
monthly charges/maintenance: _______

Other voice communication services

Main office: equipment cost: _________
installation cost per line:_________
monthly charges:_______

Satellite office: equipment cost: _________
installation cost per line: _________
monthly charges: _______

Home office: equipment cost: __________
installation cost per line: _________
monthly charges: _______

Data communications

LAN Connectivity

This section refers to the cost for LAM connectivity. Enter the costs for other desktop equipment in the "Desktop Equipment" section.

Central site

equipment (LAN CARD) cost per port:________
installation cost per port_________
support cost per month per port___________

Satellite office

Will satellite office users have LAN connectivity at both the central office and the home office?_________________

Will there be a LAN-to-LAN connection between the satellite office and the main office, or will satellite-office users individually access the main-office LAN through the public network?______________

LAN-to-LAN

Leased line charges: installation: _________
monthly charges: _________

End-user equipment (LAN card) cost per port:_________
installation cost per port:_________

Network connectivity/hardware and software (router, bridge, CSU/DSU, remote-access device):
equipment cost:_________
installation:_________
service contracts:_________

Data Security Costs (Costs incurred for secure ID or encryption software):
equipment cost:_________
installation:_________
service contracts:_________

Public Network

What type of data communication line will be used (analog, ISDN, frame relay)?_________________

installation cost per line: _________
per minute or per packet charge: _________
average minutes or packets/month/user:_________

End-user equipment (modem, TA or FRAD)cost per port:_________
installation cost per port:_________

End-user remote-access software cost per port:_________________

Central site remote-access support (Remote-access server and software):
equipment/software cost: _________
installation: _________
service contracts: _________

Data Security Costs (costs incurred for secure ID or encryption software):
equipment cost: _________
installation:_________
service contracts:_________

Home office

Will home-office users have LAN connectivity at both the central office and the home office?_______

What type of data communication line will be used (analog, ISDN, frame relay)?_______

installation cost per line: _________
per minute or per packet charge: _________
average minutes or packets/month/user:________

End-user equipment (modem, TA or FRAD)cost per port:________
installation cost per port:_________

End-user remote-access software cost per port:________________

Central site remote-access support (Remote-access server and software):
equipment/software cost: _________
installation: _________
service contracts: _________

Data Security Costs: (costs incurred for secure ID or encryption software)
equipment cost: _________
installation:_________
service contracts:_________

Desktop equipment

This section refers to non-recurring equipment costs *per user.* The costs at the central site or satellite office may refer to a percentage of a shared resource. For example, each home office many require a fax machine while users at a satellite office or main office may share a fax machine and have only a percentage of the costs allocated to them.

Central site:

Computer ____________________
Printer ____________________
Copier ____________________

Fax ____________
Application software ____________
Installation ____________

Satellite office:
Computer ____________
Printer ____________
Copier ____________
Fax ____________
Application software ____________
Installation ____________

Home Office:
Computer ____________
Printer ____________
Copier ____________
Fax ____________
Application software ____________
Installation ____________

Equipment tracking and inventory

Would your equipment tracking and inventory system need to be changed to support home office and satellite offices?_________

Estimate the additional costs for a satellite office_________
Estimate the additional costs for a home office_________

Data archiving

Will your current desktop backup and archiving strategy support a satellite office or home offices?_________

Estimate the additional costs of backing up a satellite office _________
Estimate the additional costs for backing up home offices_________

Desktop technical support questionnaire

For the purposes of this questionnaire, desktop support includes:

1. User questions regarding software applications such as Microsoft Office and e-mail.
2. Support of desktop hardware such as computers, modems, or printers.

What is the current cost per employee for providing desktop (help desk) support at your main office?______________

How will desktop (help desk) support be provided to employees working at home part time?______________

What will be the cost per employee/year?______________

How will desktop (help desk) support be provided to employees working at a satellite office?______________

What will be the cost per employee/year?______________

Appendix E: Survey Results

The following are the results of a survey we conducted to help us understand managers' views and attitudes about distributed teams. We collected responses from 514 managers at high-technology companies. The surveys were collected over a period of 18 months between July 1995 and January 1997. In reviewing the data, it is important to realize it was collected from managers who were interested enough to attend a course, trade show, or lecture on distributed teams.

Relative to colocated team members, in your opinion, remote team members are:

a)	much easier to manage	0.39%
b)	somewhat easier to manage	4.47%
c)	as easy to manage as company-located workers	6.61%
d)	somewhat more difficult to manage	58.37%
e)	much more difficult to manage	30.16%

Given a choice between two candidates with equal technical qualifications, one a remote worker and one worker who would work in your office, would you be:

a) strongly inclined to hire the remote worker	0.58%
b) somewhat inclined to hire the remote worker	17.32%
c) not inclined to consider location a factor	14.59%
d) somewhat inclined to hire a colocated worker	24.32%
e) strongly inclined to hire a colocated worker	43.19%

In your opinion, are remote team members more or less productive than colocated team members?

a) more productive	13.42%
b) less productive	39.49%
c) the same	47.08%

Relative to projects where all team members are colocated projects that use remote workers and remote work groups take:

a) much less calendar time to complete	3.31%
b) somewhat less calendar time to complete	11.28%
c) the same amount of calendar time to complete	29.18%
d) somewhat more calendar time to complete	40.47%
e) much more calendar time to complete	15.76%

On a scale of 1 to 5, how serious is the communication problem created by geographically separating workers on the same project team? (1 = not at all serious, 5 = extremely serious)

a) not at all serious	3.31%
b) somewhat less serious	11.28%
c) no difference	37.16%
d) somewhat more serious	30.35%
e) extremely serious	17.90%

When you create a project plan, how often (on average) do you schedule a deliverable from each team member?

a) daily	2.14%
b) every 1-2 weeks	47.08%
c) every 3-4 weeks	32.49%

d) every 4-8 weeks	16.93%
e) greater than 8 weeks	1.36%

How often do you communicate, either verbally or in writing, with remote team members?

a) multiple times a day	2.72%
b) daily	44.36%
c) weekly	36.96%
d) monthly	14.79%
e) I do not currently manage remote team members	1.17%

How often do you communicate with your colocated team members?

a) multiple times a day	11.67%
b) daily	33.27%
c) weekly	50.78%
d) monthly	4.28%

In comparison with face-to-face meetings, voiceconference meetings are:

a) much less effective	21.40%
b) somewhat less effective	51.75%
c) about the same	12.84%
d) more effective	0.58%
e) much more effective	10.12%
f) I have never used them	3.31%

In comparison with face-to-face meetings, videoconference meetings are:

a) much less effective	13.42%
b) somewhat less effective	34.82%
c) about the same	16.93%
d) more effective	1.17%
e) much more effective	27.04%
f) I have never used them	6.61%

About the Author

Martha Haywood is a Senior Consulting Partner at Management Strategies, a consulting firm that specializes in management of geographically distributed teams. Ms. Haywood has held director and senior management positions at high-technology companies such as Telebit Corporation and Harris Corporation. She has more than fifteen years of experience managing product development teams and extensive experience in the management of large-scale telecommunication and data communication product development projects. Ms. Haywood began her career as an electrical engineer. The combination of technical and management experience led her to specialize in the area of distributed teams. She and her partners continue to manage projects on a contract basis for companies such as Infoseek, NetFrame, and Conductus Corporation. Ms. Haywood teaches project management at the University of California Berkeley Extension and developed its course, "Managing at a Distance." At Management Strategies, Ms. Haywood has assisted companies such as Oracle, Lockheed

Martin, Hewlett Packard, and Amdahl in the training and implementation of their distributed teams. Ms. Haywood serves on the board of directors of the IEEE Engineering Management Society and the Telecommuting and Home Office Association Conference. She is a frequent lecturer at the Project World Conference, Boston University's Frontiers in Project Management Conference, ISDN World, and the PMI conference.

Index

The Artech House Technology Management and Professional Development Library

Bruce Elbert, Series Editor

Applying Total Quality Management to Systems Engineering, Joe Kasser

Designing the Networked Enterprise, Igor Hawryszkiewycz

Engineer's and Manager's Guide to Winning Proposals, Donald V. Helgeson

Evaluation of R&D Processes: Effectiveness Through Measurements, Lynn W. Ellis

Global High-Tech Marketing: An Introduction for Technical Managers and Engineers, Jules E. Kadish

Introduction to Innovation and Technology Transfer, Ian Cooke, Paul Mayes

Managing Engineers and Technical Employees: How to Attract, Motivate, and Retain Excellent People, Douglas M. Soat

Managing Virtual Teams: Practical Techniques for High-Technology Project Managers, Martha Haywood

Preparing and Delivering Effective Technical Presentations, David L. Adamy

Successful Marketing Startegy for High-Tech Firms, Eric Viardot

Successful Proposal Strategies for Small Businesses: Winning Government, Private Sector, and International Contracts, Robert S. Frey

Survival in the Software Jungle, Mark Norris

The New High-Tech Manager: Six Rules for Success in Changing Times, Kenneth Durham and Bruce Kennedy

For further information on these and other Artech House titles, including previously considered out-of-print books now available through our In-Print-Forever™ (IPF™) program, contact:

Artech House
685 Canton Street
Norwood, MA 02062
781-769-9750
Fax: 781-769-6334
Telex: 951-659
email: artech@artech-house.com

Artech House
Portland House, Stag Place
London SW1E 5XA England
+44 (0) 171-973-8077
Fax: +44 (0) 171-630-0166
Telex: 951-659
email: artech-uk@artech-house.com

Find us on the World Wide Web at:
www.artech-house.com